Contents

1	**Higher-Order Hardy-Rogers-Type Contraction Mapping Theorem**	**2**
1.1	Introduction	2
1.2	Preliminaries	2
1.3	Main Results	3
1.4	Exercises.	5
1.5	References	6
2	**Uniqueness Theorem for Best Proximity Point of Higher-Order Hardy-Rogers-Type-Wardowski Contraction**	**6**
2.1	Introduction	6
2.2	Preliminaries	6
2.3	Main Results	8
2.4	Exercises.	10
2.5	References	11
3	**Weakly Higher-Order Hardy-Rogers Type Contraction Mapping Theorem for Non-Self Maps with Partial Orders**	**11**
3.1	Introduction	11
3.2	Preliminaries	11
3.3	Main Results	13
3.4	Exercises.	16
3.5	References	17
4	**Fixed Point Theorems for Higher-Order Hardy-Rogers-Type Graphic Contraction**	**17**
4.1	Introduction	17
4.2	Preliminaries	17
4.3	Main Results	21
4.4	Exercises.	26
4.5	References	27
5	**Existence and Uniqueness of Fixed Point for α-ψ-Geraghty Higher-Order-Hardy-Rogers-Type Contraction Maps**	**28**
5.1	Introduction	28
5.2	Preliminaries	28
5.3	Main Results	30
5.4	Exercises.	33
5.5	References	34
6	**Rakotch Type Fixed Point Theorems for the Higher-Order-Hardy-Rogers Type Map using notion of w-distance**	**35**
6.1	Introduction	35
6.2	Preliminaries	35
6.3	Main Results	38
6.4	Exercises.	41
6.5	References	42

1 Higher-Order Hardy-Rogers-Type Contraction Mapping Theorem

1.1 Introduction

> **Abstract A.1 1**
>
> Let (X,d) be a metric space. Recall [Canad. Math. Bull. Vol. 16 (2), 1973] that a map $T: X \to X$ is called a Hardy-Rogers map if it satisfies $d(Tx, Ty) \leq ad(x, Tx) + bd(y, Ty) + cd(x, Ty) + ed(y, Tx) + fd(x, y)$, where a, b, c, e, f are nonnegative and satisfy $a + b + c + e + f < 1$. In the present chapter we show some techniques to give a characterization of higher-order hardy-rogers type contraction mapping theorem.

1.2 Preliminaries

> **Definition A.1 1**
>
> Let (X,d) be a metric space. A map $T: X \to X$ will be called a Hardy-Rogers type map if it satisfies, $d(Tx, Ty) \leq k[d(x, Tx) + d(y, Ty) + d(x, Ty) + d(y, Tx) + d(x, y)]$, where $k < \frac{1}{5}$ is nonnegative.

> **Definition A.2 1**
>
> Let (X,d) be a metric space and let the self-map $T: X \to X$ satisfy $d(T^r x, T^r y) \leq \sum_{q=0}^{r-1} c_q [d(T^q x, T^{q+1} x) + d(T^q y, T^{q+1} y) + d(T^q x, T^{q+1} y) + d(T^q y, T^{q+1} x) + d(T^q x, T^q y)]$ for all x, y in X and $r \in \mathbb{N}$. If $0 \leq c_q < \frac{1}{5}$ for all $0 \leq q \leq r-1$, then, $T: X \to X$ is said to be an rth order Hardy-Rogers type contraction mapping.

> **Proposition A.3 1**
>
> Let (X,d) be a metric space, and let T be an rth-order Hardy-Rogers type mapping on X. For every pair $x \neq y \in X$, define
>
> $$Z := Z(x,y) = \max_{0 \leq v \leq r-1} \beta^{-v} \frac{d(T^v x, T^v y)}{d(x, Tx) + d(y, Ty) + d(x, Ty) + d(y, Tx) + d(x, y)}$$
>
> then
>
> $$Z = \max_{n \in 0 \cup \mathbb{N}} \beta^{-n} \frac{d(T^n x, T^n y)}{d(x, Tx) + d(y, Ty) + d(x, Ty) + d(y, Tx) + d(x, y)}$$
>
> where $\beta \in [0, \frac{1}{5})$

> **Remark A.4 1**
>
> Proof of the above proposition uses same technique in obtaining Proposition 4.1 [Ezearn Fixed Point Theory and Applications (2015) 2015:88]

Definition A.5 1

Let (X,d) be a metric space, $Z \geq 1$ be the bound given by previous proposition, and $\beta \in [0, \frac{1}{5})$. The self-map $T : X \to X$ will be called an rth order Hardy-Rogers type contraction mapping if it satisfies $d(T^r x, T^r y) \leq Z\beta^r[d(x,Tx) + d(y,Ty) + d(x,Ty) + d(y,Tx) + d(x,y)]$

1.3 Main Results

Lemma A.1 1

Suppose $d(T^r x, T^r y) \leq Z\beta^r[d(x,Tx) + d(y,Ty) + d(x,Ty) + d(y,Tx) + d(x,y)]$ holds on (X,d). Since $Z\beta^r < 1$ for any $r \in \mathbb{N}$, then there exist $\gamma < 1$ such that $d(T^r x, T^{2r} x) \leq \gamma d(x, T^r x)$

Proof of Lemma A.1 1

Set $y = T^r x$ in $d(T^r x, T^r y) \leq Z\beta^r[d(x,Tx)+d(y,Ty)+d(x,Ty)+d(y,Tx)+d(x,y)]$, we obtain,

$$d(T^r x, T^{2r} x) \leq Z\beta^r[d(x, Tx) + d(T^r x, T^{r+1} x) + d(x, T^{r+1} x) + d(T^r x, Tx) \\ + d(x, T^r x)]$$
$$\leq Z\beta^r[2d(x, T^r x) + d(T^r x, T^{r+1} x) + d(x, T^{r+1} x)]$$
$$\leq Z\beta^r[2d(x, T^r x) + d(T^r x, T^{2r} x) + d(x, T^{2r} x)]$$

From the above one deduces that

$$d(T^r x, T^{2r} x) \leq \frac{2Z\beta^r}{1 - Z\beta^r} d(x, T^r x) + \frac{Z\beta^r}{1 - Z\beta^r} d(x, T^{2r} x)$$

Since by the triangle inequality, $d(T^r x, T^{2r} x) \geq d(T^{2r} x, x) - d(T^r x, x)$, it follows from the inequality immediately above that

$$d(T^{2r} x, x) - d(T^r x, x) \leq \frac{2Z\beta^r}{1 - Z\beta^r} d(x, T^r x) + \frac{Z\beta^r}{1 - Z\beta^r} d(x, T^{2r} x)$$

Upon simplifying the inequality immediately above, we obtain

$$d(T^{2r} x, x) \leq \frac{1 + Z\beta^r}{1 - 2Z\beta^r} d(x, T^r x)$$

Now using the inequality immediately above in

$$d(T^r x, T^{2r} x) \leq \frac{2Z\beta^r}{1 - Z\beta^r} d(x, T^r x) + \frac{Z\beta^r}{1 - Z\beta^r} d(x, T^{2r} x)$$

We obtain, upon simplification,

$$d(T^r x, T^{2r} x) \leq \frac{3Z\beta^r}{1 - 2Z\beta^r} d(x, T^r x)$$

Thus $\gamma := \frac{3Z\beta^r}{1 - 2Z\beta^r}$ satisfies the conclusion of the lemma.

Theorem A.2 1

Let (X, d) be a metric space, and let $T : X \mapsto X$ satisfy Definition A.5. If X is complete, then T has a unique fixed point.

Proof of Theorem A.2 1

By Lemma A.1, there exists $\frac{3Z\beta^r}{1-2Z\beta^r} := \gamma < 1$ such that $d(T^r x, T^{2r} x) \leq \gamma d(x, T^r x)$.
Let $m > n$, then,

$$d(T^{rm}x, T^{rn}x) \leq d(T^{rm}x, T^{r(m-1)}x) + \cdots + d(T^{r(n+1)}x, T^{rn}x)$$
$$\leq \gamma^n(1 + \cdots + \gamma^{m-n-1})d(x, T^r x)$$
$$\leq \frac{\gamma^n}{1-\gamma}d(x, T^r x)$$

Consequently, the sequence $\{T^{rn}x\}$ is Cauchy, and so converges to some x_0 in X. We claim $x_0 = T^r x_0$. If not, observe that

$$d(x_0, T^r x_0) \leq d(T^{r(n+1)}x, T^r x_0) + d(T^{r(n+1)}x, x_0)$$
$$\leq Z\beta^r d(T^{rn}x, T^{r(n+1)}x) + Z\beta^r d(x_0, T^r x_0) + Z\beta^r d(T^{rn}x, T^r x_0)$$
$$+ (Z\beta^r + 1)d(T^{r(n+1)}x, x_0) + Z\beta^r d(T^{rn}x, x_0)$$

Going in the limit of above inequality, we conclude that $d(x_0, T^r x_0) \leq 2Z\beta^r d(x_0, T^r x_0)$, and thus, $(1 - 2Z\beta^r)d(x_0, T^r x_0) \leq 0$. However, $1 - 2Z\beta^r > 0$ by the choice of γ, thus, $d(x_0, T^r x_0) = 0$, hence, $x_0 = T^r x_0$. Uniqueness follows from Definition A.5 and the choice of γ.

1.4 Exercises

Characterization of Reich Mapping: Exercise A.1 1

Using the definition of a Reich mapping [Canad. Math. Bull. Vol. 14 (1), 1971], show that if $T : X \mapsto X$ satisfies $d(Tx, Ty) \leq k[d(x, Tx) + d(y, Ty) + d(x, y)]$, where $k < \frac{1}{3}$ is nonnegative, then T is a Reich-Type Mapping

Characterization of Higher-Order Reich Mapping: Exercise A.2 1

Let (X, d) be a metric space and let the self-map $T : X \to X$ satisfy $d(T^r x, T^r y) \leq \sum_{q=0}^{r-1} c_q[d(T^q x, T^{q+1} x) + d(T^q y, T^{q+1} y) + d(T^q x, T^q y)]$ for all x, y in X and $r \in \mathbb{N}$, where $0 \leq c_q < \frac{1}{3}$ for all $0 \leq q \leq r - 1$. Prove that $T : X \to X$ is an rth order Reich type contraction mapping by mathematical induction on $r \in \mathbb{N}$

Alternate Characterization of Higher-Order Reich Mapping: Exercise A.3 1

Let (X, d) be a metric space. By modifying Proposition A.3, show that there exist $J \geq 1$ and $\Gamma \in [0, \frac{1}{3})$ such that, $d(T^r x, T^r y) \leq J\Gamma^r[d(x, Tx) + d(y, Ty) + d(x, y)]$, thus we have an alternate characterization of the higher-order Reich type contraction mapping

Characterization of Theorem 3, Canad. Math. Bull. Vol. 14 (1), 1971:Exercise A.4 1

Let (X, d) be a metric space. Using the alternate characterization of the higher-order Reich type contraction from the previous exercise, prove that $T : X \mapsto X$ has a unique fixed point, provided X is complete.

> **Verifying Reich-Type Mapping: Exercise A.5 1**
>
> Let $X = [0,1]$. Define $T: X \mapsto X$ by $Tx = \frac{x}{3}$ for $0 \le x < 1$ and $T(1) = \frac{1}{6}$. Prove that with $k = \frac{11}{54}$, T is a Reich type mapping.

> **Verifying Higher-Order Reich-Type Mapping: Exercise A.6 1**
>
> Let $X = [0,1]$. Define, for any $r \in \mathbb{N}$, $T: X \mapsto X$ by $T^r x = \frac{x}{3^r}$ for $0 \le x < 1$ and $T^r(1) = \frac{1}{6}$. Prove that with $\Gamma'' \in [0, \frac{11}{54})$, there exist $J'' \ge 1$ modified from Exercise A.3 such that T is a higher-order Reich type mapping.

1.5 References

(1) Canad. Math. Bull. Vol. 16 (2), 1973

(2) Ezearn Fixed Point Theory and Applications (2015) 2015:88

(3) Canad. Math. Bull. Vol. 14 (1), 1971

2 Uniqueness Theorem for Best Proximity Point of Higher-Order Hardy-Rogers-Type-Wardowski Contraction

2.1 Introduction

> **Abstract B.1 1**
>
> Let (X, d) be a metric space. Recall [Canad. Math. Bull. Vol. 16 (2), 1973] that a map $T: X \to X$ is called a Hardy-Rogers map if it satisfies $d(Tx, Ty) \le ad(x, Tx) + bd(y, Ty) + cd(x, Ty) + ed(y, Tx) + fd(x, y)$, where a, b, c, e, f are nonnegative and satisfy $a + b + c + e + f < 1$. Inspired by ideas contained in [Wardowski Fixed Point Theory and Applications 2012, 2012:94] we introduce a notion of higher-order hardy-rogers-type-wardowski contraction, and consider the map is a non-self mapping to obtain a unique best proximity point theorem.

2.2 Preliminaries

> **Definition B.1 1**
>
> Let A and B be two nonempty subsets of a metric space X. The following will be useful throughout
>
> (a) $d(y, A) := inf\{d(x,y) : x \in A\}$
>
> (b) $d(A, B) := inf\{d(x,y) : x \in A, y \in B\}$
>
> (c) $A_0 = \{x \in A : d(x,y) = d(A,B) \; for \; some \; y \in B\}$
>
> (d) $B_0 = \{y \in B : d(x,y) = d(A,B) \; for \; some \; x \in A\}$

> **Definition B.2 1**
>
> We say $x \in A$ is a best r-proximity point of the mapping $T: A \to B$ if $d(x, T^r x) = d(A, B)$ for any $r \in \mathbb{N}$

Remark B.3 1

If in Definition B.2, $r = 1$ and $A = B$, the best r-proximity point becomes a fixed point

Definition B.4 1

Let (A, B) be a pair of nonempty subsets of a metric space X with $A \neq \emptyset$. Then the pair (A, B) is said to have the P-property if and only if $d(x_1, y_1) = d(A, B)$ and $d(x_2, y_2) = d(A, B)$ implies $d(x_1, y_1) = d(x_2, y_2)$, where $x_1, x_2 \in A_0$ and $y_1, y_2 \in B_0$

Remark B.5 1

For any nonempty subset A of X, the pair (A, A) has the P-property

Definition B.6 1

Let $HOHRTW : \mathbb{R}^+ \to \mathbb{R}$ be a mapping satisfying

(a) $\alpha < \beta$ implies $HOHRTW(\alpha) < HOHRTW(\beta)$ for all $\alpha, \beta \in \mathbb{R}^+$

(b) Let $n \in \mathbb{N}$. For each sequence $\{\theta_n\}$ of positive numbers, $\lim_{n \to \infty} \theta_n = 0$ if and only if $\lim_{n \to \infty} HOHRTW(\theta_n) = -\infty$

(c) For any $r \in \mathbb{N}$, $\lim_{\theta \to 0^+} \theta^{Z\beta^r} HOHRTW(\theta) = 0$, where $Z \geq 1$ is given by Proposition A.3 and $\beta \in [0, \frac{1}{5})$

A map $T : X \mapsto X$ is said to be a HOHRTW-contraction if there exist $\tau > 0$ such that for all $x, y \in X$, $(d(T^r x, T^r y)) > 0$ implies $\tau + HOHRTW(d(T^r x, T^r y)) \leq HOHRTW(d(x, Tx) + d(y, Ty) + d(x, Ty) + d(y, Tx) + d(x, y))$

Example B.7 1

Let $HOHRTW : \mathbb{R}^+ \to \mathbb{R}$ be given by the formula $HOHRTW(\alpha) = ln(\alpha)$. Clearly, $HOHRTW$ satisfies all conditions of the previous definition. Each mapping $T : X \mapsto X$ satisfying $\tau + HOHRTW(d(T^r x, T^r y)) \leq HOHRTW(d(x, Tx) + d(y, Ty) + d(x, Ty) + d(y, Tx) + d(x, y))$ is a HOHRTW-contraction such that

$$d(T^r x, T^r y) \leq e^{-\tau}(d(x, Tx) + d(y, Ty) + d(x, Ty) + d(y, Tx) + d(x, y))$$

for all $x, y \in X$, $Tx \neq Ty$. Note that for any $r \in \mathbb{N}$ and $x, y \in X$ such that $T^r x = T^r y$, the inequality,

$$d(T^r x, T^r y) \leq e^{-\tau}(d(x, Tx) + d(y, Ty) + d(x, Ty) + d(y, Tx) + d(x, y))$$

still holds. In particular T is a higher-order Hardy-Rogers-Type contraction with $\tau = ln(\frac{1}{Z\beta^r})$

2.3 Main Results

Remark B.1 1

Let $T^r : A \mapsto B$ for any $r \in \mathbb{N}$

(a) When we write $T(A_0) \subseteq_r B_0$ we mean $T^r(A_0) \subseteq B_0$

(b) When we say T is r-continuous we mean T^r is continuous

Theorem B.2 1

Let A and B be two nonempty, closed subsets of a complete metric space X such that A_0 is nonempty. Let $T : A \mapsto B$ be an $HOHRTW$-contraction non-self mapping such that $T(A_0) \subseteq_r B_0$. Assume that the pair (A, B) has the P-property. Then there exists a unique x^* in A such that $d(x^*, T^r x^*) = d(A, B)$.

Proof of Theorem B.2 1

Choose $x_0 \in A_0$. Since $Tx_0 \in T(A_0) \subseteq_r B_0$, there exist $x_1 \in A_0$ such that $d(x_1, T^r x_0) = d(A, B)$. Again since $Tx_1 \in T(A_0) \subseteq_r B_0$, there exist $x_2 \in A_0$ such that $d(x_2, T^r x_1) = d(A, B)$. Continuing this process, we can find a sequence $\{x_n\}$ in A_0 such that $d(x_{n+1}, T^r x_n) = d(A, B)$ for all $n \in \mathbb{N}$. Since (A, B) satisfies the P-property, it follows from $d(x_{n+1}, T^r x_n) = d(A, B)$, that, $d(x_n, x_{n+1}) = d(T^r x_{n-1}, T^r x_n)$ for all $n \in \mathbb{N}$. We now show that the sequence $\{x_n\}$ is convergent in A_0. If there exists $q \in \mathbb{N}$ such that $d(T^r x_{q-1}, T^r x_q) = 0$, then since, $d(x_n, x_{n+1}) = d(T^r x_{n-1}, T^r x_n)$, it follows that $d(x_q, x_{q+1}) = 0$, which implies that $x_q = x_{q+1}$. Therefore, $T^r x_q = T^r x_{q+1}$ implies that $d(T^r x_q, T^r x_{q+1}) = 0$. Since, $T^r x_q = T^r x_{q+1}$ implies that $d(T^r x_q, T^r x_{q+1}) = 0$, and $d(x_n, x_{n+1}) = d(T^r x_{n-1}, T^r x_n)$, we deduce that $d(x_{q+2}, x_{q+1}) = d(T^r x_{q+1}, T^r x_q) = 0$ implies $x_{q+2} = x_{q+1}$. Therefore, $x_n = x_q$ for all $n \geq q$, and so $\{x_n\}$ is convergent in A_0. Now let $d(T^r x_{n-1}, T^r x_n) \neq 0$ for all $n \in \mathbb{N}$. Since T is a HOHRTW-contraction and $d(x_n, x_{n+1}) = d(T^r x_{n-1}, T^r x_n)$ holds, then for any positive integer n, we have, $\tau + HOHRTW(d(T^r x_n, T^r x_{n-1})) \leq HOHTRW(d(x_{n-1}, x_n) + d(x_{n-1}, Tx_{n-1}) + d(x_n, Tx_n) + d(x_{n-1}, Tx_n) + d(x_n, Tx_{n-1}))$. Equivalently, we have, $\tau + HOHRTW(d(x_n, x_{n+1})) \leq HOHTRW(d(x_{n-1}, x_n) + d(x_{n-1}, x_n) + d(x_n, x_{n+1}) + d(x_{n-1}, x_{n+1}))$. If $HOHRTW(\alpha) = ln(\alpha)$, then from, $\tau + HOHRTW(d(x_n, x_{n+1})) \leq HOHTRW(d(x_{n-1}, x_n) + d(x_{n-1}, x_n) + d(x_n, x_{n+1}) + d(x_{n-1}, x_{n+1}))$, one deduces upon using the triangle inequality that, $d(x_n, x_{n+1}) \leq \frac{3e^{-\tau}}{1-2e^{-\tau}} d(x_{n-1}, x_n)$, therefore, it follows that, $HOHRTW(d(x_n, x_{n+1})) \leq HOHRTW(d(x_{n-1}, x_n)) - ln(\frac{1-2e^{-\tau}}{3e^{-\tau}})$, and consequently, $HOHRTW(d(x_n, x_{n+1})) \leq HOHRTW(d(x_0, x_1)) - n\,ln(\frac{1-2e^{-\tau}}{3e^{-\tau}})$. Put $\delta_n := d(x_n, x_{n+1})$. Since, $HOHRTW(d(x_n, x_{n+1})) \leq HOHRTW(d(x_0, x_1)) - n\,ln(\frac{1-2e^{-\tau}}{3e^{-\tau}})$, it follows that, $\lim_{n \to \infty} HOHRTW(\delta_n) = -\infty$. Thus, from Definition B.6, $\lim_{n \to \infty} \delta_n = 0$. Also from Definition B.6, $\lim_{n \to \infty} \delta_n^{Z\beta^r} HOHRTW(\delta_n) = 0$, where $Z \geq 1$ is given by Proposition A.3 and $\beta \in [0, \frac{1}{5})$. It follows from, $HOHRTW(d(x_n, x_{n+1})) \leq HOHRTW(d(x_0, x_1)) - n\,ln(\frac{1-2e^{-\tau}}{3e^{-\tau}})$, that, $HOHRTW(\delta_n) - HOHRTW(\delta_0) \leq -n\,ln(\frac{1-2e^{-\tau}}{3e^{-\tau}})$, and hence, $\delta_n^{Z\beta^r} HOHRTW(\delta_n) - \delta_n^{Z\beta^r} HOHRTW(\delta_0) \leq -\delta_n^{Z\beta^r} n\,ln(\frac{1-2e^{-\tau}}{3e^{-\tau}}) \leq 0$. Thus taking limits in the inequality, $\delta_n^{Z\beta^r} HOHRTW(\delta_n) - \delta_n^{Z\beta^r} HOHRTW(\delta_0) \leq -\delta_n^{Z\beta^r} n\,ln(\frac{1-2e^{-\tau}}{3e^{-\tau}}) \leq 0$, we conclude that, $\lim_{n \to \infty} \delta_n^{Z\beta^r} n = 0$. Hence there exists a natural number $n_1 \in \mathbb{N}$ such that $\delta_n^{Z\beta^r} n \leq 1$ for all $n \geq n_1$. Therefore for any $n \geq n_1$, $\delta_n \leq \frac{1}{n^{\frac{1}{Z\beta^r}}}$, and hence, $\sum_{i=1}^{\infty} \delta_i < \infty$. Now let $m \geq n \geq n_1$, then by the triangular inequality and the fact that $\delta_n \leq \frac{1}{n^{\frac{1}{Z\beta^r}}}$, we see that $d(x_n, x_m) \leq \delta_{m-1} + \delta_{m-2} + \cdots + \delta_n \leq \sum_{i=n}^{\infty} \delta_i < \infty$, therefore, $\{x_n\}$ is Cauchy in A. Since (X, d) is complete and A is a closed subset of X, there exists x^* in A such that $\lim_{n \to \infty} x_n = x^*$. Since T is r-continuous, we have $T^r x_n \to T^r x^*$. Hence by continuity of d, $d(x_{n+1}, T^r x_n) \to d(x^*, T^r x^*)$. Since, $d(x_{n+1}, T^r x_n) = d(A, B)$, then, $d(x^*, T^r x^*) = d(A, B)$. It follows that x^* is a best r-proximity point of T. Finally, we show uniqueness of the best r-proximity point. Let $x_1, x_2 \in A$ such that $x_1 \neq x_2$ and $d(x_1, T^r x_1) = d(x_2, T^r x_2) = d(A, B)$, then by the P-property of (A, B), we have $d(x_1, x_2) = d(T^r x_1, T^r x_2)$. Also $x_1 \neq x_2$ implies $d(x_1, x_2) \neq 0$. Therefore, we have, $HOHRTW(d(T^r x_1, T^r x_2)) \leq HOHRTW(d(x_1, x_2) + d(x_1, Tx_1) + d(x_2, Tx_2) + d(x_1, Tx_2) + d(x_2, Tx_1)) - \tau$. Equivalently, we have, $HOHRTW(d(x_1, x_2)) \leq HOHRTW(3d(x_1, x_2)) - \tau$. If $HOHRTW(\alpha) = ln(\alpha)$, then from, $HOHRTW(d(x_1, x_2)) \leq HOHRTW(3d(x_1, x_2)) - \tau$

> **Proof of Theorem B.2 Continued 1**
>
> we deduce that, $d(x_1, x_2) \leq 3e^{-\tau}d(x_1, x_2)$. However, $1 - 3e^{-\tau} > 0$ and d is non-negative, thus we get the contradiction, and uniqueness follows. Also we can get the contradiction, and hence uniqueness as follows. Since $d(x_1, x_2) \leq 3e^{-\tau}d(x_1, x_2)$, it follows that $HOHRTW(d(x_1, x_2)) \leq HOHRTW(d(x_1, x_2)) - [ln(\frac{1}{3}) + \tau] < HOHRTW(d(x_1, x_2))$

If we set $A = B$ in Theorem B.2 we get the following

> **Corollary B.3 1**
>
> Let (X, d) be a complete metric space and A be a nonempty closed subset of X. Let $T : A \mapsto A$ be a $HOHRTW$-contraction self-map. Then T has a unique fixed point x^* in A

If we set $A = B$ and $HOHRTW(\alpha) = ln(\alpha)$ in Theorem B.2, we get the following

> **Corollary B.4 1**
>
> Let (X, d) be a complete metric space and A be a nonempty closed subset of X. Let $T : A \mapsto A$ be a higher-order hardy-rogers type contractive self-map. Then T has a unique fixed point x^* in A.

> **Remark B.5 1**
>
> Corollary B.4 gives Theorem A.2

2.4 Exercises

> **Exercise B.1 1**
>
> Let $GHOHRTW(\alpha) = \log_a(\alpha)$, where $a > 1$. Prove that a map T satisfying $\tau + GHOHRTW(d(T^r x, T^r y)) \leq GHOHRTW(d(x, Tx) + d(y, Ty) + d(x, Ty) + d(y, Tx) + d(x, y))$ is a higher-order Hardy-Rogers-Type contraction iff $\tau = \log_a(\frac{1}{Z\beta^\tau})$. Deduce that $GHOHRTW$-contraction is a proper extension of $HOHRTW$-contraction.

> **Exercise B.2 1**
>
> Under appropirate conditions, prove that a GHOHRTW-contraction non-self mapping has a unique best r-proximity point.

> **Exercise B.3 1**
>
> Under appropirate conditions, deduce from Exercise B.2, that a GHOHRTW-contraction self-mapping has a unique r-fixed point

> **Exercise B.4 1**
>
> Explain how one deduces the higher-order Hardy-Rogers-Type Mapping Theorem from Exercise B.1 and Exercise B.3

> **Wardowski-Type Characterization of Higher Order Reich Type Map: Exercise B.5 1**
>
> Let $GHORTW(\alpha) = log_a(\alpha)$, where $a > 1$. Prove that a map T satisfying $\tau + GHORTW(d(T^r x, T^r y)) \leq GHORTW(d(x, Tx) + d(y, Ty) + d(x, y))$ is a higher-order Reich-Type contraction (Exercise A.3) iff $\tau = log_a(3) + log_a(\frac{1}{J\Gamma^r})$. Thus, define precisely what it means to say a map $T : X \mapsto X$ is a generalized higher-order Reich-Type Wardwoski-contraction

> **Problems involving GHORTW-contractions: Exercise B.6 1**
>
> (a) Under appropriate conditions, prove that a GHORTW-contraction non-self mapping has a unique best r-proximity point.
>
> (b) Under appropirate conditions, deduce from (a), that a GHORTW-contraction self-mapping has a unique r-fixed point
>
> (c) Explain how one deduces the higher-order Reich-Type Mapping Theorem (Exercise A.4) from Exercise B.5 and (b)

2.5 References

(1) Canad. Math. Bull. Vol. 16 (2), 1973

(2) Wardowski Fixed Point Theory and Applications 2012, 2012:94

3 Weakly Higher-Order Hardy-Rogers Type Contraction Mapping Theorem for Non-Self Maps with Partial Orders

3.1 Introduction

> **Abstract C.1 1**
>
> Let (X, d) be a metric space. Recall [Canad. Math. Bull. Vol. 16 (2), 1973] that a map $T : X \to X$ is called a Hardy-Rogers map if it satisfies $d(Tx, Ty) \leq ad(x, Tx) + bd(y, Ty) + cd(x, Ty) + ed(y, Tx) + fd(x, y)$, where a, b, c, e, f are nonnegative and satisfy $a + b + c + e + f < 1$. Taking inspiration from [Harjani et al, Comput. Math. Appl. 61:790-796, 2011] we introduce a notion of weakly- higher-order hardy-rogers type contraction for non-self mappings and establish some best proximity point theorems for this class.

3.2 Preliminaries

> **Definition C.1 1**
>
> Let (X, d) be a metric space. A mapping $T : X \mapsto X$ will be called a weakly HOHRT-contraction if for all $x, y \in X$, $d(T^r x, T^r y) \leq [d(x, y) + d(x, Tx) + d(y, Ty) + d(x, Ty) + d(y, Tx)] - \psi(d(x, y), d(x, Tx), d(y, Ty), d(x, Ty), d(y, Tx))$, where $\psi : [0, \infty)^5 \mapsto [0, \infty)$ is a continuous and nondecreasing function such that $\psi(a, b, c, d, e) = 0$ if and only if $a = b = c = d = e = 0$.

Remark C.2 1

Definition A.5 follows from Definition C.1 if we take $\psi(a,b,c,d,e) = (1-Z\beta^r)(a+b+c+d+e)$

Remark C.3 1

For the remainder of this section we assume that X is a nonempty set such that $(X, \preceq)$ is a partially ordered set and (X, d) is a complete metric space.

Notation C.4 1

Let A and B be two nonempty subsets of a metric space X. The following will be useful throughout

(a) $d(A,B) := inf\{d(x,y) : x \in A, y \in B\}$

(b) $A_0 = \{x \in A : d(x,y) = d(A,B) \; for \; some \; y \in B\}$

(c) $B_0 = \{y \in B : d(x,y) = d(A,B) \; for \; some \; x \in A\}$

Definition C.5 1

A mapping $T : X \mapsto X$ is said to be r-increasing if $x \preceq y$ implies $Tx \preceq_r Ty$ for all $x, y \in X$.

Remark C.6 1

When we write $Tx \preceq_r Ty$ we mean $T^r x \preceq T^r y$ for any $r \in \mathbb{N}$

Definition C.7 1

A mapping $T : A \mapsto B$ will be called r-proximally order-preserving if and only if it satisfies the condition that $x \preceq y$, $d(u, T^r x) = d(A, B)$, $d(v, T^r y) = d(A, B)$ implies $u \preceq v$ for all $u, v, x, y \in A$ and any $r \in \mathbb{N}$

Remark C.8 1

If in Definition C.7, $A = B$, then we recover Definition C.5

Definition C.9 1

A point $x \in A$ is called a best r-proximity point of the mapping $T : A \mapsto B$ if $d(x, T^r x) = d(A, B)$ for any $r \in \mathbb{N}$

Definition C.10 1

A mapping $T : A \mapsto B$ will be called a generalized r-proximal HOHRT-contraction if for all $u, v, x, y \in A$ and any $r \in \mathbb{N}$ it satisfies $x \preceq y$, $d(u, T^r x) = d(A, B)$, $d(v, T^r y) = d(A, B)$ implies $d(u, v) \leq [d(x,y) + d(x,v) + d(y,u) + d(x,u) + d(y,v)] - \psi(d(x,y), d(x,v), d(y,u), d(x,u), d(y,v))$, where $\psi : [0, \infty)^5 \mapsto [0, \infty)$ is a continuous and nondecreasing function such that $\psi(a,b,c,d,e) = 0$ if and only if $a = b = c = d = e = 0$

3.3 Main Results

> **Theorem C.1 1**
>
> Let X be a nonempty set such that $(X, \preceq)$ is a partially ordered set and let (X, d) be a complete metric space. Let A and B be nonempty closed subsets of X such that A_0 and B_0 are nonempty. Let $T : A \mapsto B$ satisfy the following conditions
>
> (a) T is a r-continuous, r-proximally order preserving and generalized r-proximal HOHRT-contraction such that $T(A_0) \subseteq_r B_0$
>
> (b) there exists elements $x_0, x_1 \in A$ such that $x_0 \subseteq x_1$ and $d(x_1, T^r x_0) = d(A, B)$
>
> Then there exist a point $x \in A$ such that $d(x, T^r x) = d(A, B)$

Proof of Theorem C.1 1

By hypothesis (b), there exists $x_0, x_1 \in A_0$ such that $x_0 \preceq x_1$ and $d(x_1, T^r x_0) = d(A, B)$. Since $T(A_0) \subseteq_r B_0$, there exists $x_2 \in A_0$ such that $d(x_2, T^r x_1) = d(A, B)$. Since T is r-proximally order preserving it follows that $x_1 \preceq x_2$. Continuing this process, we can find a sequence $\{x_n\}$ in A_0 such that $x_{n-1} \preceq x_n$ and $d(x_n, T^r x_{n-1}) = d(A, B)$. Having found x_n, we can choose a point $x_{n+1} \in A_0$ such that $x_n \preceq x_{n+1}$ and $d(x_n, T^r x_{n+1}) = d(A, B)$. Since T is a generalized r-proximal HOHRT-contraction, for each $n \in \mathbb{N}$, we have,

$$d(x_n, x_{n+1}) \leq [d(x_{n-1}, x_n) + d(x_{n-1}, x_{n+1}) + d(x_n, x_n) + d(x_{n-1}, x_n) + d(x_n, x_{n+1})]$$
$$- \psi(d(x_{n-1}, x_n), d(x_{n-1}, x_{n+1}), d(x_n, x_n), d(x_{n-1}, x_n), d(x_n, x_{n+1}))$$
$$= [d(x_{n-1}, x_n) + d(x_{n-1}, x_{n+1}) + d(x_{n-1}, x_n) + d(x_n, x_{n+1})]$$
$$- \psi(d(x_{n-1}, x_n), d(x_{n-1}, x_{n+1}), 0, d(x_{n-1}, x_n), d(x_n, x_{n+1}))$$
$$\leq 2d(x_{n-1}, x_n) + d(x_{n-1}, x_{n+1}) + d(x_n, x_{n+1})$$
$$\leq 3d(x_{n-1}, x_n) + 2d(x_n, x_{n+1})$$
$$\leq \frac{1}{3} d(x_{n-1}, x_n) + \frac{2}{3} d(x_n, x_{n+1})$$

From the expression immediately above we find that $d(x_n, x_{n+1}) \leq d(x_{n-1}, x_n)$, thus, $\{d(x_n, x_{n+1})\}$ is non-increasing and bounded below. So there exists $w \geq 0$ such that $\lim_{n \to \infty} d(x_n, x_{n+1}) = w$. Thus,

$$w \leq \lim_{n \to \infty} [\frac{1}{3} d(x_{n-1}, x_n) + \frac{2}{3} d(x_n, x_{n+1})]$$
$$\leq \frac{1}{3} w + \frac{2}{3} w$$
$$= w$$

It follows that $\lim_{n \to \infty} [d(x_{n-1}, x_n) + 2d(x_n, x_{n+1})] = 3w$. Now observe that,

$$3w = \lim_{n \to \infty} [d(x_{n-1}, x_n) + 2d(x_n, x_{n+1})]$$
$$\leq \lim_{n \to \infty} [3d(x_{n-1}, x_{n+1})] + \lim_{n \to \infty} [2d(x_{n-1}, x_n) + d(x_n, x_{n+1})]$$
$$\leq \lim_{n \to \infty} [3d(x_{n-1}, x_{n+1})] + 3w$$

Thus, $\lim_{n \to \infty} d(x_{n-1}, x_{n+1}) = 0$. Now consider inequality statement below

$$d(x_n, x_{n+1}) \leq [d(x_{n-1}, x_n) + d(x_{n-1}, x_{n+1}) + d(x_{n-1}, x_n) + d(x_n, x_{n+1})]$$
$$- \psi(d(x_{n-1}, x_n), d(x_{n-1}, x_{n+1}), 0, d(x_{n-1}, x_n), d(x_n, x_{n+1}))$$
$$\leq \frac{1}{3} d(x_{n-1}, x_n) + \frac{2}{3} d(x_n, x_{n+1})$$

Proof of Theorem C.1 Continued 1

Taking limits in the inequality statement immediately above, we get the following

$$w \leq 3w - \psi(w,0,0,w,w) \leq \frac{1}{3}w + \frac{2}{3}w = w$$

From the inequality immediately above, it follows that $3w - \psi(w,0,0,w,w) = w$ or $\psi(w,0,0,w,w) = 2w$. However, by property of ψ, $\psi(w,0,0,w,w) = 0$ if and only if $2w = 0$ if and only if $w = 0$. It follows that $\lim_{n\to\infty} d(x_n, x_{n+1}) = 0$. Now we show that the sequence $\{x_n\}$ is Cauchy. Suppose that $\{x_n\}$ is not a Cauchy sequence. Then there exists $\epsilon > 0$ and subsequences $\{x_{m_k}\}, \{x_{n_k}\}$ of $\{x_n\}$ such that $n_k > m_k \geq k$ with $r_k := d(x_{m_k}, x_{n_k}) \geq \epsilon$ and $d(x_{m_k}, x_{n_k-1}) < \epsilon$ for each $k \in \mathbb{N}$. For each $n \geq 1$, let $\alpha_n := d(x_{n+1}, x_n)$. It follows that,

$$\epsilon \leq r_k$$
$$\leq d(x_{m_k}, x_{n_k-1}) + d(x_{n_k-1}, x_{n_k})$$
$$< \epsilon + \alpha_{n_k-1}$$

Since $\lim_{n\to\infty} d(x_n, x_{n+1}) = 0$, it follows from the inequality immediately above that $\lim_{k\to\infty} r_k = \epsilon$. Now observe that

$$r_k = d(x_{m_k}, x_{n_k})$$
$$\leq d(x_{n_k}, x_{m_k+1}) + d(x_{m_k+1}, x_{m_k})$$
$$= d(x_{n_k}, x_{m_k+1}) + \alpha_{m_k}$$
$$\leq d(x_{n_k}, x_{m_k}) + d(x_{m_k}, x_{m_k+1}) + \alpha_{m_k}$$
$$= r_k + \alpha_{m_k} + \alpha_{m_k}$$

Since $\lim_{k\to\infty} \alpha_{m_k} = 0$, it follows from the inequality immediately above that $\epsilon = \lim_{k\to\infty} r_k = \lim_{k\to\infty} d(x_{n_k}, x_{m_k+1})$. Similarly, we can show that $\epsilon = \lim_{k\to\infty} d(x_{m_k}, x_{n_k+1})$. By construction of $\{x_n\}$ we may assume that $x_{m_k} \preceq x_{n_k}$ such that $d(x_{n_k+1}, T^r x_{n_k}) = d(A, B)$ and $d(x_{m_k+1}, T^r x_{m_k}) = d(A, B)$. It follows upon using the triangle inequality and the fact that T is a generalized r-proximal HOHRT-contraction, we have,

$$\epsilon \leq r_k$$
$$\leq d(x_{m_k}, x_{m_k+1}) + d(x_{n_k}, x_{n_k+1}) + d(x_{m_k+1}, x_{n_k+1})$$
$$= \alpha_{m_k} + \alpha_{n_k} + d(x_{m_k+1}, x_{n_k+1})$$
$$\leq \alpha_{m_k} + \alpha_{n_k} + d(x_{m_k}, x_{n_k}) + d(x_{m_k}, x_{n_k+1}) + d(x_{n_k}, x_{m_k+1})$$
$$+ d(x_{m_k}, x_{m_k+1}) + d(x_{n_k}, x_{n_k+1})$$
$$- \psi(d(x_{m_k}, x_{n_k}), d(x_{m_k}, x_{n_k+1}), d(x_{n_k}, x_{m_k+1}), d(x_{m_k}, x_{m_k+1}), d(x_{n_k}, x_{n_k+1}))$$

Take limit in the inequality immediately above, we get that $\epsilon \leq 3\epsilon - \psi(\epsilon, \epsilon, \epsilon, 0, 0) \leq \epsilon$. Thus, $\psi(\epsilon, \epsilon, \epsilon, 0, 0) = 2\epsilon$ and by the property of ψ, $\psi(\epsilon, \epsilon, \epsilon, 0, 0) = 0$ if and only if $2\epsilon = 0$ if and only if $\epsilon = 0$, which is a contradiction. Thus $\{x_n\}$ is Cauchy. Since A is a closed subset of the complete metric space X, there exists $x \in A$ such that $\lim_{n\to\infty} x_n = x$. Since T is r-continuous, that is, T^r is continuous, and $\lim_{n\to\infty} x_n = x$, if we take limits in $d(x_n, T^r x_{n+1}) = d(A, B)$, it follows that $d(x, T^r x) = d(A, B)$.

If we take $\psi(a, b, c, d, e) = (1 - Z\beta^r)(a + b + c + d + e)$, then $\psi(a, b, c, d, e) = 0$ if and only if $a = b = c = d = e = 0$, thus from the theorem immediately above. We get the following

Corollary C.2 1

Let X be a nonempty set such that $(X, \preceq)$ is a partially ordered set and let (X, d) be a complete metric space. Let A and B be nonempty closed subsets of X such that A_0 and B_0 are nonempty. Let $T : A \mapsto B$ satisfy the following conditions

(a) T is r-continuous, r-increasing such that $T(A_0) \subseteq_r B_0$ and $x \preceq y$, $d(u, T^r x) = d(A, B)$ and $d(v, T^r y) = d(A, B)$ implies that $d(u, v) \leq Z\beta^r [d(x, y) + d(x, v) + d(y, u) + d(x, u) + d(y, v)]$, where $Z \geq 1$ and $\beta \in [0, \frac{1}{5})$ are given by Proposition A.3

(b) there exist $x_0, x_1 \in A_0$ such that $x_0 \preceq x_1$ and $d(x_1, T^r x_0) = d(A, B)$

Then there exists a point $x \in A$ such that $d(x, T^r x) = d(A, B)$

3.4 Exercises

Exercise C.1 1

Verify Remark C.2

Exercise C.2 1

Using Remark C.6, verify Remark C.8

Exercise C.3 1

For any $r \in \mathbb{N}$, $\beta \in [0, \frac{1}{5})$, and $Z \geq 1$, let $\psi(a, b, c, d, e) = (1 - Z\beta^r)(a + b + c + d + e)$, where $\psi : [0, \infty)^5 \mapsto [0, \infty)$ is continuous and non-decreasing. Prove that, $\psi(a, b, c, d, e) = 0$ if and only if $a = b = c = d = e = 0$.

Exercise C.4 1

Let (X, d) be a metric space. Suppose a map $T : X \mapsto X$ satisfies for all $x, y \in X$, $d(T^r x, T^r y) \leq [d(x, y) + d(x, Tx) + d(y, Ty)] - \psi(d(x, y), d(x, Tx), d(y, Ty))$, where $\psi : [0, \infty)^3 \mapsto [0, \infty)$ is a continuous and nondecreasing function such that $\psi(a, b, c) = 0$ if and only if $a = b = c = 0$. Prove that T is a higher-order Reich type contraction (Exercise A.3) iff $\psi(a, b, c) = (1 - J\Gamma^r)(a + b + c)$. Thus, define the concept of a weakly higher-order Reich type contraction

Exercise C.5 1

Let X be a nonempty set such that $(X, \preceq)$ is a partially ordered set and let (X, d) be a metric space. Let A and B be nonempty closed subsets of X. Suppose $T : A \mapsto B$ satisfies for all $u, v, u', v', x, y \in A$ and any $r \in \mathbb{N}$, $x \preceq y$, $d(u, T^r x) = d(A, B)$, $d(v, T^r y) = d(A, B)$, $d(u', Tx) = d(A, B)$, $d(v', Ty) = d(A, B)$, implies $d(u, v) \leq [d(x, y) + d(x, u') + d(y, v')] - \psi(d(x, y), d(x, u'), d(y, v'))$, where $\psi : [0, \infty)^3 \mapsto [0, \infty)$ is a continuous and nondecreasing function such that $\psi(a, b, c) = 0$ if and only if $a = b = c = 0$. Prove that if the underlying map is a self-map, then it is a higher order Reich type contraction (Exercise A.3) iff $\psi(a, b, c) = (1 - J\Gamma^r)(a+b+c)$. Thus, define the concept of a generalized r-proximal higher-order Reich type contraction

> **Exercise C.6 1**
>
> Prove the following or give a counterexample to show it is false: Theorem C.1 holds if generalized r-proximal HOHRT-contraction is replaced with generalized r-proximal higher-order Reich type contraction (Exercise C.5)

> **Exercise C.7 1**
>
> Deduce from Exercise C.6 that Theorem C.1 holds if the continuous nondecreasing function is given by $\psi(a,b,c) = (1 - J\Gamma^r)(a+b+c)$, where $J \geq 1$ and Γ come from Exercise A.3. Otherwise, give a counterexample showing we cannot get such a Corollary from Exercise C.6

3.5 References

(1) Canad. Math. Bull. Vol. 16 (2), 1973

(2) Harjani et al, Comput. Math. Appl. 61:790-796, 2011

4 Fixed Point Theorems for Higher-Order Hardy-Rogers-Type Graphic Contraction

4.1 Introduction

> **Abstract D.1 1**
>
> Let (X,d) be a metric space. Recall [Canad. Math. Bull. Vol. 16 (2), 1973] that a map $T: X \to X$ is called a Hardy-Rogers map if it satisfies $d(Tx, Ty) \leq ad(x, Tx) + bd(y, Ty) + cd(x, Ty) + ed(y, Tx) + fd(x, y)$, where a, b, c, e, f are nonnegative and satisfy $a+b+c+e+f < 1$. Taking inspiration from Jachymski [J. Jachymski, The contraction principle for mappings on a metric space endowed with a graph, Proc. Amer. Math. Soc., 136(2008), 1359-1373] we prove some fixed point theorems for graphic contractions of the higher-order-hardy-rogers type using comparison function and also study existence and uniqueness in complete metric spaces with a graph.

4.2 Preliminaries

> **Remark D.1 1**
>
> Recall that a map f from a metric space (X,d) into itself is called a weakly Picard operator (WPO) if $\lim_{n \to \infty} f^n x = z$ for all $x \in X$, and z is a fixed point of f. Moreover, the fixed point of f is unique, and we say that f is a Picard operator (PO).

Higher-order contraction mapping principle was given in [Ezearn Fixed Point Theory and Applications (2015) 2015:88] and an alternate characterization was given in [Ampadu, Clement (2015): Generalization of Higher Order Contraction Mapping Theorem. Unpublished]. It follows from these papers and the above remark, that we have the following.

Definition D.2 1

A map f from a metric space (X,d) into itself will be called a weakly r-Picard operator (r-WPO) if $\lim_{n\to\infty} f^{rn}x = z$ for all $x \in X$ and any $r \in \mathbb{N}$, and z is a r-fixed point of f, that is, $f^r z = z$ for any $r \in \mathbb{N}$. Moreover, the r-fixed point of f is unique, and we say that f is a r-Picard operator (r-PO).

Notation D.3 1

Let (X,d) be a metric space. Δ will denote the diagonal of the Cartesian product $X \times X$

Notation D.4 1

Let G be a directed graph. $V(G)$ will denote the set of vertices of the graph and $E(G)$ will denote the set of edges of the graph.

Remark D.5 1

We assume $V(G)$ coincides with X. We also assume that $E(G)$ contains all loops, that is, $E(G) \supseteq \Delta$.

Remark D.6 1

If G has no parallel edges, one can identify G with the pair $(V(G), E(G))$

Definition D.7 1

The conversion of the graph G is denoted by G^{-1}, which is a graph obtained from G by reversing the direction of the edges. In particular, $E(G^{-1}) = \{(y,x) \in X \times X : (x,y) \in E(G)\}$

Notation D.8 1

$\widetilde{G}$ will denote the undirected graph obtained from G by omitting the direction of the edges.

Remark D.9 1

If $\widetilde{G}$ is a directed graph for which $E(\widetilde{G})$ is symmetric, then $E(\widetilde{G}) = E(G) \cup E(G^{-1})$

Definition D.10 1

If $x, y \in V' \subseteq V$ and $(x,y) \in E' \subseteq E$, then (V', E') is called a subgraph of G

Definition D.11 1

If x and y are vertices in a graph G, then a path from x to y of length $N \in \mathbb{N}$ is a sequence $\{x_i\}_{i=0}^{N}$ of $N+1$ vertices such that $x_0 = x$, $x_N = y$ and $(x_{n-1}, x_n) \in E(G)$ for $i = 1, \cdots, N$

Definition D.12 1

Let G be a graph

(a) G is connected if there is a path between any two vertices

(b) G is weakly connected if $\widetilde{G}$ is connected

Definition D.13 1

If G is such that $E(G)$ is symmetric and x is a vertex in G, then the subgraph G_x consisting of all edges and vertices which are contained in some path beginning at x is called the component of G containing x. In particular, $V(G) = [x]_G$ where $[x]_G$ denotes the equivalence class of relation $\Re$ defined on $V(G)$ by the rule: $y\Re z$ if there is a path in G from y to z

Definition D.14 1

If $f : X \mapsto X$ is an operator, then we denote by $F_r(f) = \{x \in X : x = f^r x\}$ for any $r \in \mathbb{N}$, the set of all r-fixed points of f

Definition D.15 1

For any $r \in \mathbb{N}$, $X_r^f = \{x \in X : (x, f^r x) \in E(G) \text{ or } (f^r x, x) \in E(G)\}$

Definition D.16 1

A mapping $f : X \mapsto X$ will be called a *G-Higher-Order-Hardy-Rogers-Type* contraction if the following conditions hold

(a) f, r-preserves edges of G, that is, $(x, y) \in E(G) \Rightarrow (f^r x, f^r y) \in E(G)$ for all $x, y \in X$ and any $r \in \mathbb{N}$

(b) f, r-decreases weights of edges of G, that is, $(x, y) \in E(G) \Rightarrow d(f^r x, f^r y) \leq Z\beta^r[d(x, fx) + d(y, fy) + d(x, fy) + d(y, fx) + d(x, y)]$ for all $x, y \in X$ and any $r \in \mathbb{N}$, where $Z \geq 1$ and $\beta \in [0, \frac{1}{5})$ are given by Proposition A.3

Definition D.17 1

A mapping $f : X \mapsto X$ will be called a *G-Graphic-Higher-Order-Hardy-Rogers-Type* contraction if the following conditions hold

(a) f, r-preserves edges of G, that is, $(x, y) \in E(G) \Rightarrow (f^r x, f^r y) \in E(G)$ for all $x, y \in X$ and any $r \in \mathbb{N}$

(b) f, r-decreases weights of edges of G, that is, $(x, y) \in E(G) \Rightarrow d(f^r x, f^{2r} x) \leq Z\beta^r[d(x, fx) + d(f^r x, f^{r+1} x) + d(x, f^{r+1} x) + d(f^r x, fx) + d(x, f^r x)]$ for all $x \in X_r^f$ and any $r \in \mathbb{N}$, where $Z \geq 1$ and $\beta \in [0, \frac{1}{5})$ are given by Proposition A.3

Definition D.18 1

A map $f : X \mapsto X$ will be called r-orbitally continuous if for all $x, y \in X$, $r \in \mathbb{N}$, and any sequence $\{k_n\}_{n \in \mathbb{N}}$ of positive integers, $f^{rk_n} x \to y \Rightarrow f^r(f^{rk_n} x) \to f^r y$ as $n \to \infty$

Definition D.19 1

A map $f : X \mapsto X$ will be called r-orbitally G-continuous if for all $x, y \in X$, $r \in \mathbb{N}$, and any sequence $\{k_n\}_{n \in \mathbb{N}}$ of positive integers, $f^{rk_n}x \to y, (f^{rk_n}x, f^{r(k_{n+1})}x) \in E(G) \Rightarrow f^r(f^{rk_n}x) \to f^r y$ as $n \to \infty$

Definition D.20 1

A map $\varphi : \mathbb{R}^+ \mapsto \mathbb{R}^+$ is called a comparison function if it satisfies the following

(a) $t_1 \leq t_2 \Rightarrow \varphi(t_1) \leq \varphi(t_2)$

(b) $\{\varphi^n(t)\}_{n \in \mathbb{N}}$ converges to 0 for all $t > 0$

Remark D.21 1

If $\varphi : \mathbb{R}^+ \mapsto \mathbb{R}^+$ is a comparison function, then $\varphi(t) < t$, for all $t > 0$, $\varphi(0) = 0$ and φ is right continuous at 0

Definition D.22 1

A mapping $f : X \mapsto X$ will be called a (G, φ)-Higher-Order-Hardy-Rogers-Type contraction if the following conditions hold

(a) f, r-preserves edges of G, that is, $(x, y) \in E(G) \Rightarrow (f^r x, f^r y) \in E(G)$ for all $x, y \in X$ and any $r \in \mathbb{N}$

(b) there exist a comparison function $\varphi : \mathbb{R}^+ \mapsto \mathbb{R}^+$ such that $d(f^r x, f^r y) \leq \varphi(d(x, fx) + d(y, fy) + d(x, fy) + d(y, fx) + d(x, y))$ for all $(x, y) \in E(G)$ and any $r \in \mathbb{N}$

Definition D.23 1

[S.M.A Aleomraninejad, Sh. Rezapour, N. Shahzad, Some fixed point results on a metric space with a graph, Topology and Its Applications, (159), (2012), 659-663] The graph G is called a (C)-graph whenever for each sequence $\{x_n\}_{n \geq 0}$ in X with $x_n \to x$ and $(x_n, x_{n+1}) \in E(G)$ for all $n \geq 0$, there is a subsequence $\{x_{n_k}\}_{k \geq 0}$ such that $(x_{n_k}, x) \in E(G)$ for all $k \geq 0$

Definition D.24 1

Let (X, d) be a metric space. The operator $f : X \mapsto X$ will be called a Higher-Order Hardy-Rogers-Type map if it satisfies Definition A.5

Definition D.25 1

A mapping $f : X \mapsto X$ will be called a (G, φ)-Graphic-Higher-Order-Hardy-Rogers-Type contraction if the following conditions hold

(a) f, r-preserves edges of G, that is, $(x, y) \in E(G) \Rightarrow (f^r x, f^r y) \in E(G)$ for all $x, y \in X$ and any $r \in \mathbb{N}$

(b) there exist a comparison function $\varphi : \mathbb{R}^+ \mapsto \mathbb{R}^+$ such that $d(f^r x, f^{2r} x) \leq \varphi(d(x, fx) + d(f^r x, f^{r+1} x) + d(x, f^{r+1} x) + d(f^r x, fx) + d(x, f^r x))$ for all $x \in X_r^f$ and any $r \in \mathbb{N}$

Remark D.26 1

If f is a (G,φ)-Graphic-Higher-Order-Hardy-Rogers-Type contraction, then f is both a (G^{-1},φ)-Graphic-Higher-Order-Hardy-Rogers-Type contraction and a $(\widetilde{G},\varphi)$-Graphic-Higher-Order-Hardy-Rogers-Type contraction

Example D.27 1

Any G-Graphic-Higher-Order-Hardy-Rogers-Type contraction is a (G,φ)-Graphic-Higher-Order-Hardy-Rogers-Type contraction if the comparison function, $\varphi:\mathbb{R}^+ \mapsto \mathbb{R}^+$, is given by $\varphi(t) = Z\beta^r t$, where $Z \geq 1$ and $\beta \in [0,\frac{1}{5})$ are given by Proposition A.3

Notation D.28 1

$X_f^r = \{x \in X : (x, f^r x) \in E(G)\}$

Definition D.29 1

A mapping $f : X \mapsto X$ will be called a Higher-Order-Hardy-Rogers-Type G-contraction if the following conditions hold

(a) f, r-preserves edges of G, that is, $(x,y) \in E(G) \Rightarrow (f^r x, f^r y) \in E(G)$ for all $x, y \in X$ and any $r \in \mathbb{N}$

(b) $d(f^r x, f^r y) \leq Z\beta^r(d(x, fx) + d(y, fy) + d(x, fy) + d(y, fx) + d(x, y))$ for all $(x,y) \in E(G)$ and any $r \in \mathbb{N}$, where $Z \geq 1$ and $\beta \in [0,\frac{1}{5})$ are given by Proposition A.3

Remark D.30 1

If f is a Higher-Order-Hardy-Rogers-Type G-contraction, then f is both a Higher-Order-Hardy-Rogers-Type G^{-1}-contraction and a Higher-Order-Hardy-Rogers-Type $\widetilde{G}$-contraction

Example D.31 1

Let $X = \{0, 1, 2, 3\}$ and $d(x, y) = |x - y|$ for all $x, y \in X$. Define $f : X \mapsto X$ by $f^r x = 0$, if $x \in \{0, 1\}$ and $f^r x = 1$, if $x \in \{2, 3\}$, for any $r \in \mathbb{N}$. Let $E(G) = \{(0,1), (0,2), (2,3), (0,0), (1,1), (2,2), (3,3)\}$, then $(f^r x, f^r y) \in E(G)$, whenever $(x, y) \in E(G)$. Let $\sigma \in [0, \frac{4}{25})$ and $Q \geq 1$ be obtained by modifying Proposition A.3, then $d(f^r x, f^r y) \leq Q\sigma^r(d(x, fx) + d(y, fy) + d(x, fy) + d(x, y))$ for all $(x, y) \in E(G)$, and f is a (modified) higher-order hardy rogers type G-contraction.

4.3 Main Results

Lemma D.1 1

Let (X, d) be a metric space endowed with a graph G. Let $f : X \mapsto X$ be a (G, φ)-Graphic-Higher-Order-Hardy-Rogers-Type contraction. If $x \in X_r^f$, then there exist $q(x) \geq 0$ such that $d(f^{rn}x, f^{r(n+1)}x) \leq \varphi^n(q(x))$ for all $n \in \mathbb{N}$

Proof of Lemma D.1 1

Take $x \in X_r^f$, that is, $(x, f^r x) \in E(G)$ or $(f^r x, x) \in E(G)$. If $(x, f^r x) \in E(G)$, notice that since f is a (G, φ)-Graphic-Higher-Order-Hardy-Rogers-Type contraction it follows that

$$d(f^r x, f^{2r} x) \leq \varphi(d(x, f^r x) + d(f^r x, f^{r+1} x) + d(x, f^{r+1} x) + d(f^r x, fx) + d(x, f^r x))$$
$$\leq \varphi(d(x, f^r x) + d(x, f^{r+1} x) + d(x, f^r x))$$
$$\leq \varphi(3 d(x, f^r x))$$

So by induction we have $(f^{rn} x, f^{r(n+1)} x) \in E(G)$. In particular,

$$d(f^{rn} x, f^{r(n+1)} x) \leq \varphi(3 d(f^{r(n-1)} x, f^{rn} x))$$
$$\leq \varphi^2(3^2 d(f^{r(n-2)} x, f^{r(n-1)} x))$$
$$\vdots$$
$$\leq \varphi^n(3^n d(x, f^r x))$$

If $(f^r x, x) \in E(G)$, then since d is symmetric, we find that $d(f^{r(n+1)} x, f^{rn} x) \leq \varphi^n(3^n d(f^r x, x))$. Thus, we arrive at the conclusion of the lemma by setting $q(x) := 3^n d(x, f^r x)$

Lemma D.2 1

Let (X, d) be a complete metric space endowed with a graph G. Assume that $f : X \mapsto X$ is a (G, φ)-graphic Higher-Order-Hardy-Rogers-Type contraction. Then for each $x \in X_r^f$, there exist $x^* \in X$ such that the sequence $\{f^{rn} x\}_{n \in \mathbb{N}}$ converges to x^* as $n \to \infty$

Proof of Lemma D.2 1

Let $n, m \in \mathbb{N}$, $m > n$, using property of φ and previous lemma, we have,

$$d(f^{rn} x, f^{r(n+m)} x) \leq d(f^{rn} x, f^{r(n+1)} x) + \cdots + d(f^{r(n+m-1)} x, f^{r(n+m)} x)$$
$$\leq \varphi^n(q(x)) + \cdots + \varphi^{(n+m-1)}(q(x))$$
$$= \sum_{j=1}^{m} \varphi^{n+j-1}(q(x)) < \infty \qquad (n, m \to \infty)$$

It follows that $\{f^{rn} x\}_{n \in \mathbb{N}}$ is Cauchy. Since (X, d) is complete, $\{f^{rn} x\}_{n \in \mathbb{N}}$ is a convergent sequence whose limit is say $x^* \in X$

Lemma D.3 1

Let (X, d) be a complete metric space endowed with a graph G, $f : X \mapsto X$ be a (G, φ)-graphic Higher-Order-Hardy-Rogers-Type contraction for which there exists $x_0 \in X$ such that $f^r x_0 \in [x_0]_{\widetilde{G}}$, for any $r \in \mathbb{N}$. Let $\widetilde{G}_{x_0}$ be the component of $\widetilde{G}$ containing x_0. Then $[x_0]_{\widetilde{G}}$ is f^r-invariant and $f^r|_{[x_0]_{\widetilde{G}}}$ is a $(\widetilde{G}_{x_0}, \varphi)$-graphic Higher-Order-Hardy-Rogers-Type contraction

Remark D.4 1

The above lemma can be proven via techniques used in Bojor[Annals of the University of Craiova, Mathematics and Computer Science Series Volume 37(4), 2010, Pages 85-92] to obtain Proposition 2.1

Remark D.5 1

We will say the triple (X, d, G) has the condition $(\star)$ if it satisfies the following: For any $\{x_n\}_{n \in \mathbb{N}}$ in X if $x_n \to x$ and $(x_n, x_{n+1}) \in E(G)$ or $(x_{n+1}, x_n) \in E(G)$ for all $n \in \mathbb{N}$, then there is a subsequence $\{x_{n_k}\}_{k \in \mathbb{N}}$ with $(x_{n_k}, x) \in E(G)$ or $(x, x_{n_k}) \in E(G)$ for all $k \in \mathbb{N}$

Theorem D.6 1

Let (X, d) be a complete metric space and G be a directed graph. Let the triple (X, d, G) have the condition$(\star)$. Let $f : X \mapsto X$ be a (G, φ)-graphic Higher-Order-Hardy-Rogers-Type contraction which is r-orbitally G-continuous. Then the following statements are equivalent

(a) $F_r(f) \neq \emptyset$ iff $X_r^f \neq \emptyset$

(b) If $X_r^f \neq \emptyset$ and G is weakly connected, then f is an r-WPO

(c) For any $x \in X_r^f$, we have, $f|_{[x]_{\widetilde{G}}}$ is a r-WPO

Proof of Theorem D.6 1

Beginning with (c). Let $x \in X_r^f$. Hence, there exist $q(x) \geq 0$ such that

$$d(f^{rn}x, f^{r(n+1)}x) \leq (Z\beta^r)^n q(x)$$

for all $n \in \mathbb{N}$, where $Z \geq 1$ and $\beta \in [0, \frac{1}{5})$ is given by Proposition A.3. So there exists $x^* \in X$ such that $\lim_{n \to \infty} f^{rn}x = x^*$. Since $x \in X_r^f$, then $f^{rn}x \in X_r^f$ for every $n \in \mathbb{N}$. Now assume that $(x, f^r x) \in E(G)$ or $(f^r x, x) \in E(G)$. By using condition $(\star)$, there exists a subsequence $\{f^{rn_k}x\}_{k \in \mathbb{N}}$ of $\{f^{rn}x\}_{n \in \mathbb{N}}$ such that $(f^{rn_k}x, x^*) \in E(G)$ or $(x^*, f^{rn_k}x,) \in E(G)$ for each $n \in \mathbb{N}$. Notice that a path in G can be formed using the points $x, f^r x, \cdots, f^{rn_1}x, x^*$, and hence $x^* \in [x]_{\widetilde{G}}$. Since f is r-orbitally G-continuous, we obtain that x^* is a fixed point for $f^r|_{[x]_{\widetilde{G}}}$. To prove (a), using (c) we have $F_r(f) \neq \emptyset$. By using the assumption that $\triangle \subseteq E(G)$, we immediately obtain that $X_r^f \neq \emptyset$. Hence (a) holds. To prove (b), let $x \in X_r^f$. By weak connectivity of G, we have that $X = [x]_{\widetilde{G}}$ and by applying (c), we obtain the desired result.

Lemma D.7 1

Let (X, d) be a metric space endowed with a graph G. Let $f : X \mapsto X$ be a higher-order hardy-rogers type G-contraction. If $x \in X_f^r$ then there exists $j(x) \geq 0$ and $\zeta < 1$ such that $d(f^{rn}x, f^{r(n+1)}x) \leq \zeta^n j(x)$ for all $n \in \mathbb{N}$

Proof of Lemma D.7 1

Take $x \in X_f^r$, then, $(x, f^r x) \in$

Proof of Lemma D.7 Continued 1

$E(G)$, then by induction we have $(f^{rn}x, f^{r(n+1)}x) \in E(G)$ for each $n \in \mathbb{N}$. Now,

$$d(f^{rn}x, f^{r(n+1)}x) \leq Z\beta^r(d(f^{r(n-1)}x, f^{rn}x) + d(f^{rn}x, f^{r(n+1)}x) + d(f^{r(n-1)}x, f^{r(n+1)}x)$$
$$+ d(f^{rn}x, f^{rn}x) + d(f^{r(n-1)}x, f^{rn}x))$$
$$= Z\beta^r(d(f^{r(n-1)}x, f^{rn}x) + d(f^{rn}x, f^{r(n+1)}x)$$
$$+ d(f^{r(n-1)}x, f^{r(n+1)}x) + d(f^{r(n-1)}x, f^{rn}x))$$
$$\leq Z\beta^r(3d(f^{r(n-1)}x, f^{rn}x) + 2d(f^{rn}x, f^{r(n+1)}x))$$

It follows from the expression immediately above that

$$d(f^{rn}x, f^{r(n+1)}x) \leq \frac{3Z\beta^r}{1 - 2Z\beta^r} d(f^{r(n-1)}x, f^{rn}x)$$

Thus, with $\zeta := \frac{3Z\beta^r}{1-2Z\beta^r} < 1$, we obtain $d(f^{rn}x, f^{r(n+1)}x) \leq \zeta^n d(x, f^r x) = \zeta^n j(x)$. Thus we reach the conclusion of the Lemma by setting $\zeta := \frac{3Z\beta^r}{1-2Z\beta^r}$ and $j(x) := d(x, f^r x)$.

Lemma D.8 1

Let (X, d) be a complete metric space endowed with a graph G. Suppose that $f : X \mapsto X$ is a higher-order hardy-rogers type G-contraction. Then for each $x \in X_f^r$, there exists $x^* \in X$ such that the sequence $\{f^{rn}x\}_{n \in \mathbb{N}}$ converges to $x^* \in X$ as $n \to \infty$

Proof of Lemma D.8 1

If $x \in X_f^r$, then $f^r x \in [x]_{\widetilde{G}}$ and $(f^{rn}, f^{r(n+1)}x) \in E(G)$ for all $n \in \mathbb{N}$. Let $n, m \in \mathbb{N}$, $m > n$, using previous lemma, we have,

$$d(f^{rn}x, f^{rm}x) \leq d(f^{rn}x, f^{r(n+1)}x) + d(f^{r(n+1)}x, f^{r(n+2)}x) + \cdots + d(f^{r(m-1)}x, f^{rm}x)$$
$$\leq \zeta^n j(x) + \zeta^{n+1} j(x) + \cdots + \zeta^{m-1} j(x)$$
$$= \zeta^n [1 + \zeta + \zeta^2 + \cdots + \zeta^{m-n-1}] j(x)$$
$$\leq \frac{\zeta^n}{1 - \zeta} j(x) \to 0 \qquad (n \to \infty)$$

Consequently, $\{f^{rn}x\}_{n \in \mathbb{N}}$ is a Cauchy sequence. By completeness of X, $\{f^{rn}x\}_{n \in \mathbb{N}}$ is a convergent sequence whose limit is say $x^* \in X$

Remark D.9 1

Lemma D.3 still holds if we replace (G, φ)-graphic Higher-Order-Hardy-Rogers-Type contraction with higher-order hardy-rogers type G-contraction and $(\widetilde{G}_{x_0}, \varphi)$-graphic Higher-Order-Hardy-Rogers-Type contraction with higher-order-hardy-rogers type $\widetilde{G}_{x_0}$-contraction

Theorem D.10 1

Let (X, d) be a complete metric space endowed with a graph G and f be a self-map on X. Assume that the following assertions hold:

(a) G is weakly connected and a (C)-graph

(b) f is a higher-order hardy-rogers type $\widetilde{G}$-contraction

(c) X_f^r is nonempty

Then f is a r-PO

Proof of Theorem D.10 1

Let $x \in X_f^r$, then $f^r x \in [x]_{\widetilde{G}}$ and $(f^{rn}x, f^{r(n+1)}x) \in E(G)$ for all $n \in \mathbb{N}$. From Lemma D.8, we get that $\{f^{rn}x\}_{n \in \mathbb{N}}$ converges to $x^* \in X$. Now we show that x^* is a r-fixed point of f. As $f^{rn}x \to x^*$, $(f^{rn}x, f^{r(n+1)}x) \in E(G)$ for all $n \in \mathbb{N}$, and G is a (C)-graph, there exists a subsequence $\{f^{rn_k}x\}$ of $\{f^{rn}x\}$ such that $(f^{rn_k}x, x^*) \in E(G)$ for each $k \in \mathbb{N}$. Since $(f^{rn_k}x, x^*) \in E(\widetilde{G})$ and f is a higher-order hardy-rogers type $\widetilde{G}$-contraction, it follows that,

$$d(f^{r(n_k+1)}x, f^r x^*) \leq Z\beta^r[d(f^{rn_k}x, f^{r(n_k+1)}x) + d(x^*, f^r x^*) + d(f^{rn_k}x, f^r x^*) \\ + d(x^*, f^{r(n_k+1)}x) + d(f^{rn_k}x, x^*)]$$

Taking limits in the above inequality as $k \to \infty$, we get that $d(x^*, f^r x^*) \leq 2Z\beta^r d(x^*, f^r x^*)$. By the choice of ζ from Lemma D.8 or Lemma D.7, it follows that $x^* = f^r x^*$. Thus, x^* is the r-fixed point of f. Now we show uniqueness of the r-fixed point. Suppose that $y^* \neq x^*$ is another r-fixed point of f. Since G is a (C)-graph, then there exists a subsequence $\{f^{rn_k}x\}$ of $\{f^{rn}x\}$ such that $(f^{rn_k}x, x^*) \in E(G)$ and $(f^{rn_k}x, y^*) \in E(G)$ for each $k \in \mathbb{N}$. Furthermore, G is weakly connected and so $(x^*, y^*) \in E(\widetilde{G})$. Thus, we have

$$d(x^*, y^*) \leq Z\beta^r[d(x^*, x^*) + d(y^*, y^*) + 3d(x^*, y^*)] \\ = 3Z\beta^r d(x^*, y^*)$$

By the choice of ζ from Lemma D.8 or Lemma D.7, it follows from the above inequality, that $x^* = y^*$, and uniqueness follows.

Remark D.11 1

The above theorem still holds if we replace the condition that G is a C-graph with r-orbitally G-continuity of f

Example D.12 1

Let $X = [0,1]$ and $d(x,y) = |x-y|$, for all $x,y \in X$. Let $E(G) = \{(x,y) : x,y \in [0,1]\}$. Define $f : X \mapsto X$ by $f^r x = \frac{x}{4^r}$, $x \in X$, and any $r \in \mathbb{N}$. We observe the following

(a) G is weakly connected and a (C)-graph

(b) X_f^r is nonempty

(c) f is r-orbitally G-continuous

(d) $(f^r x, f^r y) \in E(\widetilde{G})$ whenever $(x,y) \in E(\widetilde{G})$

Example D.12 Continued 1

(e) there exists $W \geq 1$ and $\xi \in [0, \frac{1}{6})$ modified from Proposition A.3, such that,
$d(f^r x, f^r y) \leq W\xi^r[d(x,fx) + d(y,fy) + d(x,y)]$, whenever $(x,y) \in E(\widetilde{G})$

Thus all conditions of Theorem D.10 and the Theorem obtained from Remark D.11 are satisfied. Note that 0 is the r-fixed point of f.

4.4 Exercises

Exercise D.1 1

If T is a higher-order Reich type mapping, explain how one deduces from Exercise A.4, that T is a r-WPO, and hence a r-PO ?

Exercise D.2 1

Let X, d, and f be defined as in Example D.12. Put $V(G) = X$ and $E(G) = X \times X$, then the contraction from Exercise A.4 is a higher-order Hardy-Rogers Type $\widetilde{G}$-contraction by Theorem D.10. So from Exercise A.4 we can deduce that the operator is a $r - WPO$, and hence a r-PO. Thus generalizing this example (special case of previous exercise), define precisely what it will mean for the operator in Exercise A.4 to be a higher-order Reich type $\widetilde{G}$-contraction. Further, $\widetilde{G}$ is obtained from G by omitting direction of its edges, therefore, define precisely what it will mean for the operator in Exercise A.4 to be a higher-order Reich type G-contraction.

Exercise D.3 1

As a consequence of the previous exercise, prove that Theorem D.6 is valid, if the map there is a (G,φ)- graphic higher-order Reich type contraction which is r-orbitally G-continuous. Also prove Theorem D.10 is valid, if f is replaced with the precise definition of higher-order Reich type G-contraction.

Exercise D.4 1

Verify Example D.27

Exercise D.5 1

Prove that the following is a consequence of Theorem D.10: Let (X, d) be a complete metric space endowed with a partial ordering "$\leq$" such that every pair of elements of X has an upper and a lower bound. Let $f : X \mapsto X$ be continuous and monotone, and such that $x \leq y \Rightarrow d(f^r x, f^r y) \leq Z\beta^r[d(x, fx) + d(y, fy) + d(x, fy) + d(y, fx) + d(x, y)]$ for all $x, y \in X$ and any $r \in \mathbb{N}$, where $Z \geq 1$ and $\beta \in [0, \frac{1}{5})$ are given by Proposition A.3. If there exists $x_0 \in X$ with $x_0 \leq f^r x_0$ or $f^r x_0 \leq x_0$, then f is a r-PO. This problem is a higher-order Hardy-Rogers Type characterization of a result due to Ran and Reurings in [Proc. Amer. Math. Soc. 132 (2004), 14351443]

Exercise D.6 1

Prove that the following is a consequence of Exercise D.3: Let (X, d) be a complete metric space endowed with a partial ordering "$\leq$" such that every pair of elements of X has an upper and a lower bound. Let $f : X \mapsto X$ be continuous and monotone, and such that $x \leq y \Rightarrow d(f^r x, f^r y) \leq J\Gamma^r[d(x, fx) + d(y, fy) + d(x, y)]$ for all $x, y \in X$ and any $r \in \mathbb{N}$, where $J \geq 1$ and $\Gamma \in [0, \frac{1}{3})$ are given by Exercise A.3. If there exists $x_0 \in X$ with $x_0 \leq f^r x_0$ or $f^r x_0 \leq x_0$, then f is a r-PO. This problem is a higher-order Reich Type characterization of a result due to Ran and Reurings in [Proc. Amer. Math. Soc. 132 (2004), 14351443]

4.5 References

(1) Canad. Math. Bull. Vol. 16 (2), 1973

(2) J. Jachymski, The contraction principle for mappings on a metric space endowed with a graph, Proc. Amer. Math. Soc., 136(2008) 1359-1373

(3) Ezearn Fixed Point Theory and Applications (2015) 2015:88

(4) Ampadu, Clement (2015): Generalization of Higher Order Contraction Mapping Theorem. Unpublished

(5) S.M.A Aleomraninejad, Sh. Rezapour, N. Shahzad, Some fixed point results on a metric space with a graph, Topology and Its Applications, (159), (2012), 659-663

(6) F. Bojor, Annals of the University of Craiova, Mathematics and Computer Science Series Volume 37(4), 2010, Pages 85-92

(7) Ran and Reurings, Proc. Amer. Math. Soc. 132 (2004), 14351443

5 Existence and Uniqueness of Fixed Point for α-ψ-Geraghty Higher-Order-Hardy-Rogers-Type Contraction Maps

5.1 Introduction

> **Abstract E.1 1**
>
> Let (X, d) be a metric space. Recall[Canad. Math. Bull. Vol. 16 (2), 1973] that a map $T : X \to X$ is called a Hardy-Rogers map if it satisfies $d(Tx, Ty) \leq ad(x, Tx) + bd(y, Ty) + cd(x, Ty) + ed(y, Tx) + fd(x, y)$, where a, b, c, e, f are nonnegative and satisfy $a + b + c + e + f < 1$. Taking inspiration from Samet et.al [B. Samet, C. Vetro, P. Vetro, Fixed point theorems for α-ψ-contractive type mappings, Nonlinear Analysis 75(2012) 2154-2165] and Geraghty [M. Geraghty, On contractive mappings, Proc. Amer. Math. Soc. 40(1973) 604-608] we introduce a notion of α-ψ-Geraghty higher-order Hardy-Rogers type contraction and prove existence and uniqueness type theorems in complete metric spaces.

5.2 Preliminaries

Among many generalizations of the Banach contraction principle, the result of Geraghty [M. Geraghty, On contractive mappings, Proc. Amer. Math. Soc. 40(1973) 604-608] is interesting, and it can be stated in the following way

> **Theorem E.1(a) 1**
>
> Let (X, d) be a complete metric space, and $T : X \mapsto X$ be an operator. If T satisfies the inequality: $d(Tx, Ty) \leq \beta(d(x, y))d(x, y)$, for any $x, y \in X$, where $\beta : [0, \infty) \mapsto [0, 1)$ satisfies the condition that $\beta(t_n) \to 1$ implies $t_n \to 0$, then T has a unique fixed point.

Note that if $T : X \mapsto X$ satisfies Definition A.1, then we have the following Geraghty-type characterization of the Hardy-Rogers Mapping Theorem [Theorem 1(a), Canad. Math. Bull. Vol. 16 (2), 1973] which we state without proof.

> **Theorem E.2(a) 1**
>
> Let (X, d) be a complete metric space, and $T : X \mapsto X$ be an operator. If T satisfies the inequality: $d(Tx, Ty) \leq \Omega(d(x, Tx) + d(y, Ty) + d(x, Ty) + d(y, Tx) + d(x, y))[d(x, Tx) + d(y, Ty) + d(x, Ty) + d(y, Tx) + d(x, y)]$, for any $x, y \in X$, where $\Omega : [0, \infty) \mapsto [0, \frac{1}{5})$ satisfies the condition that $\Omega(t_n) \to \frac{1}{5}$ implies $t_n \to 0$, then T has a unique fixed point.

Note that if $T : X \mapsto X$ satisfies Definition A.5, then we have the following Geraghty-type characterization of the Higher-Order Hardy-Rogers Mapping Theorem (Theorem A.2) which we state without proof.

> **Theorem E.3(a) 1**
>
> Let (X, d) be a complete metric space, and $T : X \mapsto X$ be an operator. If T satisfies the inequality: $d(T^r x, T^r y) \leq \Xi(d(x, Tx) + d(y, Ty) + d(x, Ty) + d(y, Tx) + d(x, y))[d(x, Tx) + d(y, Ty) + d(x, Ty) + d(y, Tx) + d(x, y)]$, for any $x, y \in X$, where $\Xi : [0, \infty) \mapsto [0, 1)$ satisfies the condition that $\Xi(t_n) \to 1$ implies $t_n \to 0$, then T has a unique fixed point.

Definition E.4 1

Let $T : X \mapsto X$ be a map and $\alpha : X \times X \mapsto \mathbb{R}$ be a function. We will say T is r-α-admissible if $\alpha(x, y) \geq 1$ implies $\alpha(T^r x, T^r y) \geq 1$

Example E.5 1

Let $X = \mathbb{R}$, $f^r x = x^{\frac{1}{3r}}$ for any $r \in \mathbb{N}$, and $\alpha(x, y) = e^{x-y}$. If $\alpha(x, y) = e^{x-y} \geq 1$, then $x \geq y$, which implies that $f^r x \geq f^r y$, that is, $\alpha(f^r x, f^r y) \geq 1$. So f is r-α-admissible.

Definition E.6 1

A map $T : X \mapsto X$ will be called triangular r-α-admissible if in addition to Definition E.4 it satisfies $\alpha(x, z) \geq 1$ and $\alpha(z, y) \geq 1$ implies $\alpha(x, y) \geq 1$

Example E.7 1

Let $X = \mathbb{R}$, $f^r x = x^{\frac{1}{3r}}$ for any $r \in \mathbb{N}$, and $\alpha(x, y) = e^{x-y}$. By Example E.5, f is r-α-admissible. Now notice that $\alpha(x, z) \geq 1$ and $\alpha(z, y) \geq 1$ implies $\alpha(x, z) = e^{x-z} \geq 1$ and $\alpha(z, y) = e^{z-y} \geq 1$, which implies that $x - z \geq 0$ and $z - y \geq 0$, and thus, $x - y \geq 0$, that is, $\alpha(x, y) = e^{x-y} \geq e^0 = 1$

Notation E.8 1

Ψ will denote the class of functions $\psi : [0, \infty) \mapsto [0, \infty)$ satisfying the following conditions

(a) ψ is non-decreasing

(b) ψ is subadditive, that is, $\psi(s + t) \leq \psi(s) + \psi(t)$

(c) ψ is continuous

(d) $\psi(t) = 0 \iff t = 0$

Definition E.9 1

Let (X, d) be a metric space, and let $\alpha : X \times X \mapsto \mathbb{R}$ be a function. The map $T : X \mapsto X$ will be called a generalized r-α-ψ-Geraghty HOHRT-contraction type if there exist β as given in Theorem E.1(a) such that

$$\alpha(x, y)\psi(d(T^r x, T^r y)) \leq \beta(\psi(M(x, y)))\psi(M(x, y)) \text{ for any } x, y \in X$$

where $M(x, y) := d(x, Tx) + d(y, Ty) + d(x, Ty) + d(y, Tx) + d(x, y)$ and $\psi \in \Psi$

Remark E.10 1

If in Definition E.9, $\psi(t) = t$, then T will be called a r-α-Geraghty HOHRT-contraction type

Remark E.11 1

Since β is the function defined in Theorem E.1(a), then,

$$\alpha(x,y)\psi(d(T^r x, T^r y)) \leq \beta(\psi(M(x,y)))\psi(M(x,y)) < \psi(M(x,y))$$

for any $x, y \in X$ and $x \neq y$

5.3 Main Results

Lemma E.1 1

Let $T : X \mapsto X$ be a triangular r-α-admissible map. Assume there exists $x_1 \in X$ such that $\alpha(x_1, T^r x_1) \geq 1$ for any $r \in \mathbb{N}$. Define a sequence $\{x_n\}$ by $x_{n+1} = T^r x_n$. Then, we have $\alpha(x_m, x_n) \geq 1$ for all $m, n \in \mathbb{N}$ with $m < n$.

Proof of Lemma E.1 1

Since there exist $x_1 \in X$ such that $\alpha(x_1, T^r x_1) \geq 1$ and T is triangular r-α-admissible, then $\alpha(x_2, x_3) = \alpha(T^r x_1, T^{2r} x_1) \geq 1$. Continuing, we have, $\alpha(x_n, x_{n+1}) \geq 1$ for all $n \in \mathbb{N}$. Suppose that $m < n$. Since T is triangular r-α-admissible, then $\alpha(x_m, x_{m+1}) \geq 1$ and $\alpha(x_{m+1}, x_{m+2}) \geq 1$ implies that $\alpha(x_m, x_{m+2}) \geq 1$. Similarly, $\alpha(x_m, x_{m+2}) \geq 1$ and $\alpha(x_{m+2}, x_{m+3}) \geq 1$ implies that $\alpha(x_m, x_{m+3}) \geq 1$. Continuing, we get $\alpha(x_m, x_n) \geq 1$

Theorem E.2(b) 1

Let (X, d) be a complete metric space, $\alpha : X \times X \mapsto \mathbb{R}$ be a function, and let $T : X \mapsto X$ be a map. Suppose that the following conditions are satisfied

(a) T is a generalized r-α-ψ-Geraghty HOHRT contraction type map

(b) T is triangular r-α-admissible

(c) there exists $x_1 \in X$ such that $\alpha(x_1, T^r x_1) \geq 1$

(d) T is r-continuous

Then, T has a r-fixed point $x^* \in X$ and $\{T^{rn} x_1\}$ converges to x^*

Proof of Theorem E.2(b) 1

Let $x_1 \in X$ be such that $\alpha(x_1, T^r x_1) \geq 1$. Define a sequence $\{x_n\} \subseteq X$ by $x_{n+1} = T^r x_n$ for $n \in \mathbb{N}$. Suppose that $x_{n_0} = x_{n_0+1}$ for some $n_0 \in \mathbb{N}$. It follows that x_{n_0} is an r-fixed point of T and we are done. So we assume that $x_n \neq x_{n+1}$ for all $n \in \mathbb{N}$. By Lemma E.1, $\alpha(x_n, x_{n+1}) \geq 1$ for all $n \in \mathbb{N}$. By Definition E.9, we have,

$$\psi(d(x_{n+1}, x_{n+2})) = \psi(d(T^r x_n, T^r x_{n+1}))$$
$$\leq \alpha(x_n, x_{n+1}) \psi(d(T^r x_n, T^r x_{n+1}))$$
$$\leq \beta(\psi(M(x_n, x_{n+1}))) \psi(M(x_n, x_{n+1}))$$

for all $n \in \mathbb{N}$, where

$$M(x_n, x_{n+1}) = d(x_n, Tx_{n-1}) + d(x_{n+1}, Tx_n) + d(x_n, Tx_n) + d(x_{n+1}, Tx_{n-1})$$
$$+ d(x_n, x_{n+1}) = 3d(x_n, x_{n+1})$$

Proof of Theorem E.2(b) Continued 1

Thus it follows that

$$\psi(d(x_{n+1}, x_{n+2})) \leq \beta(\psi(3d(x_n, x_{n+1}))\psi(3d(x_n, x_{n+1}))$$
$$< \psi(3d(x_n, x_{n+1}))$$
$$\leq \psi(\frac{1}{3}d(x_{n+1}, x_{n+2}))$$
$$\leq \psi(\frac{2}{3}d(x_n, x_{n+1}))$$
$$= \psi(d(x_n, x_{n+1}) - \frac{1}{3}d(x_n, x_{n+1}))$$
$$\leq \psi(d(x_n, x_{n+1})) - \psi(\frac{1}{3}d(x_n, x_{n+1}))$$
$$< \psi(d(x_n, x_{n+1}))$$

Consequently the sequence $\{d(x_n, x_{n+1})\}$ is decreasing and bounded below by some $r \geq 0$. We claim $\lim_{n \to \infty} d(x_n, x_{n+1}) = r$ and $r = 0$. If not, let $r > 0$, and observe that,

$$\frac{\psi(d(x_{n+1}, x_{n+2}))}{\psi(d(x_n, x_{n+1}))} < \frac{\psi(d(x_{n+1}, x_{n+2}))}{\psi(3d(x_n, x_{n+1}))}$$
$$\leq \beta(\psi(3d(x_n, x_{n+1})))$$
$$< 1$$

Taking limits in the above inequality, we conclude that $\lim_{n \to \infty} \beta(\psi(3d(x_n, x_{n+1}))) = 1$, thus, $\lim_{n \to \infty} \psi(3d(x_n, x_{n+1})) = 0$, which implies that $3 \lim_{n \to \infty} d(x_n, x_{n+1}) = 0$, thus, $\lim_{n \to \infty} d(x_n, x_{n+1}) = 0$, that is $r = 0$. Now observe that, $\lim_{n,m \to \infty} M(x_m, x_n) = 3 \lim_{n,m \to \infty} d(x_m, x_n)$. Now we show that $\{x_n\}$ is Cauchy. If not, observe that $\epsilon = \lim_{n,m \to \infty} \sup d(x_m, x_n) > 0$. By the triangular inequality, we have, $d(x_n, x_m) \leq d(x_n, x_{n+1}) + d(x_{n+1}, x_{m+1}) + d(x_{m+1}, x_m)$. It follows that we have the following

$$\psi(d(x_n, x_m)) \leq \psi(d(x_n, x_{n+1}) + d(T^r x_n, T^r x_m) + d(x_{m+1}, x_m))$$
$$\leq \psi(d(x_n, x_{n+1})) + \psi(d(T^r x_n, T^r x_m)) + \psi(d(x_{m+1}, x_m))$$
$$\leq \psi(d(x_n, x_{n+1})) + \beta(\psi(M(x_n, x_m)))\psi(M(x_n, x_m)) + \psi(d(x_{m+1}, x_m))$$

Since $\lim_{n \to \infty} d(x_n, x_{n+1}) = 0$, taking limits in the above inequality, we get that

$$1 = \frac{\lim_{n,m \to \infty} \psi(d(x_n, x_m))}{\lim_{n,m \to \infty} \psi(d(x_n, x_m))}$$
$$\leq \frac{\lim_{n,m \to \infty} \psi(d(x_n, x_m))}{\lim_{n,m \to \infty} \psi(3d(x_n, x_m))}$$
$$\leq \lim_{n,m \to \infty} \beta(\psi(3d(x_n, x_m)))$$

Thus, $\lim_{n,m \to \infty} \beta(\psi(3d(x_n, x_m))) = 1$. Consequently, $\lim_{n,m \to \infty} \psi(3d(x_n, x_m)) = 0$, and thus, $\lim_{n,m \to \infty} 3d(x_n, x_m) = 0$, that is, $\lim_{n,m \to \infty} d(x_n, x_m) = 0$, which is a contradiction. So the sequence $\{x_n\}$ is Cauchy. Since X is complete, there exists $x^* = \lim_{n \to \infty} x_n \in X$. Since T is r-continuous, then, $\lim_{n \to \infty} x_n = T^r x^*$, thus $x^* = T^r x^*$

Theorem E.3(b) 1

Let ψ be the identity in previous theorem, and assume for all $x, y \in Fix(T^r)$, where $Fix(T^r)$ denotes the set of r-fixed point of T, there exists $z \in X$ such that $\alpha(x, z) \geq 1$ and $\alpha(y, z) \geq 1$, then x^* is the unique r-fixed point of T

Proof of Theorem E.3(b) 1

By previous theorem, $x^* \in X$ is an r-fixed point of T. Let $y^* \in X$ be another r-fixed point of T. Now by assumption, there exist $z \in X$ such that $\alpha(x^*, z) \geq 1$ and $\alpha(y^*, z) \geq 1$. Since T is r-α-admissible, we have, $\alpha(x^*, T^{rn}z) \geq 1$ and $\alpha(y^*, T^{rn}z) \geq 1$, for all natural numbers n. It follows that

$$d(x^*, T^{rn}z) \leq \alpha(x^*, T^{r(n-1)}z) d(T^r x^*, T^r T^{r(n-1)}z)$$
$$\leq \beta(d(x^*, T^{r(n-1)}z)) d(x^*, T^{r(n-1)}z)$$
$$< d(x^*, T^{r(n-1)}z)$$

for all natural numbers n. Thus the sequence $\{d(x^*, T^{r(n-1)}z)\}$ is decreasing and bounded below by some $u \geq 0$. We claim that $\lim_{n \to \infty} d(x^*, T^{r(n-1)}z) = u$, and $u = 0$. If not observe that,

$$\frac{d(x^*, T^{rn}z)}{d(x^*, T^{r(n-1)}z)} \leq \beta(d(x^*, T^{r(n-1)}z))$$

Taking limits in the above inequality, we get $\lim_{n \to \infty} \beta(d(x^*, T^{r(n-1)}z)) = 1$, which implies that $\lim_{n \to \infty} d(x^*, T^{r(n-1)}z) = 0$, a contradiction. Thus, $u = 0$. Similarly, we have $\lim_{n \to \infty} d(y^*, T^{r(n-1)}z) = 0$. Consequently, $x^* = y^*$

5.4 Exercises

Exercise E.1 1

Recall Geraghty's characterization of the Banach Contraction Principle is given by Theorem E.1(a) . If $T : X \mapsto X$ is the map from Exercise A.1, then what is Geraghty's characterization of the Reich Mapping Theorem [Theorem 3, Canad. Math. Bull. Vol. 14 (1), 1971]?

Exercise E.2 1

Repeat the previous exercise, if $T : X \mapsto X$ is the map from Exercise A.3

> **Exercise E.3 1**
>
> Let $X = [1, \infty)$. Define $\alpha : X \times X \mapsto [0, \infty)$ by
> $$\alpha(x, y) = \begin{cases} 2 & x < y \\ \frac{1}{3} & otherwise \end{cases}$$
>
> For any $r \in \mathbb{N}$, define $f : X \mapsto X$ by
> $$f^r x = \begin{cases} \frac{1}{x} & r_{odd} \\ x & r_{even} \end{cases}$$
>
> Show that f is not r-α-admissible if r is odd, but is r-α-admissible if r is even. By comparing with Example 3.2 (arXiv:1306.3498 [math.FA]) deduce that the notion of r-α-admissible is more general than the notion of α-admissible.

> **Exercise E.4 1**
>
> Using the map in the previous exercise, show that f is not **triangular** r-α-admissible if r is odd, but is **triangular** r-α-admissible if r is even. Deduce that the notion of triangular r-α-admissible is more general than the notion of triangular α-admissible.

> **Exercise E.5 1**
>
> Let (X, d) be a complete metric space, $\alpha : X \times X \mapsto \mathbb{R}$ be a function, and $\psi \in \Psi$. Deduce from Exercise E.2, what it will mean to say that $T : X \mapsto X$ is a generalized r-α-ψ-Geraghty higher-order-Reich-type-contraction type map

> **Exercise E.6 1**
>
> Prove that Theorem E.2(b) holds if $T : X \mapsto X$ is the map in the previous exercise

> **Exercise E.7 1**
>
> If $T : X \mapsto X$ is a r-α-Geraghty higher-order Reich-type-contraction type, that is, $\psi(t) = t$ in the previous exercise, then show that the r-fixed point of the previous exercise is unique. Can we guarantee uniqueness of the r-fixed point in the previous exercise for general $\psi \in \Psi$?

5.5 References

(1) Canad. Math. Bull. Vol. 16 (2), 1973

(2) B. Samet, C. Vetro, P. Vetro, Fixed point theorems for α-ψ-contractive type mappings, Nonlinear Analysis 75(2012) 2154-2165

(3) M. Geraghty, On contractive mappings, Proc. Amer. Math. Soc. 40(1973) 604-608

(4) Canad. Math. Bull. Vol. 14 (1), 1971

(5) arXiv:1306.3498 [math.FA]

6 Rakotch Type Fixed Point Theorems for the Higher-Order-Hardy-Rogers Type Map using notion of w-distance

> **Abstract F.1 1**
>
> ## 6.1 Introduction
>
> Taking inspiration from [Rakotch, E. (1962) "A note on contractive mappings", Proc. A.M.S. 13: 459–465] and [Kada, O.; Suzuki, T.; Takahashi, W. (1996) "Non convex minimization theorems and fixed point theorems in complete metric spaces", Math. Japon. 44: 381–391] we introduce a Rakotch type characterization of the Hardy-Rogers Map using the notion of w-distance and obtain some fixed point theorems.

6.2 Preliminaries

Rakotch's problem [Rakotch, E. A note on contractive mappings. Proc. Amer. Math. Soc. 13 1962 459–465] consist of defining a family of functions $F = \{\alpha(x,y)\}$ satisfying $0 \leq \alpha(x,y) < 1$, $\sup \alpha(x,y) = 1$ and such that Banach's theorem holds when the contraction factor is replaced with any $\alpha(x,y) \in F$.

If in Theorem E.2(a), $\Omega(d(x,Tx) + d(y,Ty) + d(x,Ty) + d(y,Tx) + d(x,y)) := k$, where $k < \frac{1}{5}$ is nonnegative, then we have the following characterization of the Hardy-Rogers Mapping Theorem [Theorem 1(a), Canad. Math. Bull. Vol. 16 (2), 1973]

> **Theorem F.1(a) 1**
>
> Let (X,d) be a complete metric space, and $T : X \mapsto X$ be an operator. If T satisfies the inequality: $d(Tx,Ty) \leq k[d(x,Tx) + d(y,Ty) + d(x,Ty) + d(y,Tx) + d(x,y)]$, for any $x,y \in X$, where $k < \frac{1}{5}$ is nonnegative, then T has a unique fixed point.

Thus we have the following Rakotch type problem: Define a family of functions $\Theta = \{\heartsuit(x,y)\}$ satisfying $0 \leq \heartsuit(x,y) < \frac{1}{5}$, $\sup \heartsuit(x,y) = \frac{1}{5}$ and such that Theorem F.1(a) holds when the contraction factor is replaced with any $\heartsuit(x,y) \in \Theta$

If in Theorem E.3(a), $\Xi(d(x,Tx) + d(y,Ty) + d(x,Ty) + d(y,Tx) + d(x,y)) = Z\beta^r$, where $Z \geq 1$ and $\beta \in [0, \frac{1}{5})$ are given by Proposition A.3, then we get Theorem A.2

Thus in the case of Theorem A.2, the Rakotch type problem is to define a family of functions $\Gamma = \{\spadesuit(x,y)\}$ satisfying $0 \leq \spadesuit(x,y) < 1$, $\sup \spadesuit(x,y) = 1$ and such that Theorem A.2 holds when the contraction factor is replaced with any $\spadesuit(x,y) \in \Gamma$

Definition F.2 1

[Kada, O.; Suzuki, T.; Takahashi, W. (1996) "Non convex minimization theorems and fixed point theorems in complete metric spaces", Math. Japon. 44: 381–391]
Let (X, d) be a metric space. A function $\rho : X \times X \mapsto [0, \infty)$ is called a $w-distance$ on X if the following conditions are satisfied

(a) $\rho(x, z) \leq \rho(x, y) + \rho(y, z)$ for any $x, y, z \in X$

(b) For any $x \in X$, $\rho(x, \cdot) : X \mapsto [0, \infty)$ is lower semi continuous

(c) For any $\epsilon > 0$, there exist $\delta_\epsilon > 0$ such that $\rho(x, z) \leq \delta_\epsilon$ and $\rho(z, y) \leq \delta_\epsilon$ imply $\rho(x, y) \leq \epsilon$

Example F.3 1

The metric d is a w-distance on X

Remark F.4 1

Some further examples of w-distance are contained in [Kada, O.; Suzuki, T.; Takahashi, W. (1996) "Non convex minimization theorems and fixed point theorems in complete metric spaces", Math. Japon. 44: 381–391; Suzuki, T.; Takahashi, W. (1996) "Fixed point theorems and characterizations of metric completeness", Top. Meth. in Nonlinear Analysis 8: 371–382]

Lemma F.5 1

[Kada, O.; Suzuki, T.; Takahashi, W. (1996) "Non convex minimization theorems and fixed point theorems in complete metric spaces", Math. Japon. 44: 381–391]
Let (X, d) be a metric space and let ρ be a w-distance on X. Let $\{\alpha_n\}$ and $\{\beta_n\}$ be sequences in $[0, \infty)$ converging to zero, and let $x, y, z \in X$. Then the following hold

(a) If $\rho(x_n, y) \leq \alpha_n$ and $\rho(x_n, z) \leq \beta_n$ for any $n \in \mathbb{N}$, then $y = z$. In particular if $\rho(x, y) = 0$ and $\rho(x, z) = 0$, then $y = z$

(b) If $\rho(x_n, y_n) \leq \alpha_n$ and $\rho(x_n, z) \leq \beta_n$ for any $n \in \mathbb{N}$, then $\{y_n\}$ converges to z

(c) If $\rho(x_n, x_m) \leq \alpha_n$ for any $n, m \in \mathbb{N}$ with $m > n$ then $\{x_n\}$ is a Cauchy sequence

(d) If $\rho(y, x_n) \leq \alpha_n$ for any $n \in \mathbb{N}$, then $\{x_n\}$ is a Cauchy sequence

Definition F.6 1

Let (X, d) be a metric space. A finite sequence $\{x_0, x_1, \cdots, x_n\}$ of points of X is called an ϵ-chain joining x_0 and x_n if $d(x_{i-1}, x_i) < \epsilon$ for each $\epsilon > 0$, $i = 1, 2, \cdots, n$

Definition F.7 1

A metric space (X, d) is said to be ϵ-chainable if for each pair (x, y) of its points there exist an ϵ-chain joining x and y

Lemma F.8 1

[Suzuki, T.; Takahashi, W. (1996) "Fixed point theorems and characterizations of metric completeness", Top. Meth. in Nonlinear Analysis 8: 371–382] Let $\epsilon \in (0, \infty)$ and let (X, d) be an ϵ-chainable metric space. Then the function $\rho : X \times X \mapsto [0, \infty)$ defined by $p(x, y) = \inf\{\sum_{i=1}^{n} d(x_{i-1}, x_i) : \{x_0, x_1, \cdots, x_n\}$ is an ϵ-chain joining x and $y\}$ is a w-distance on X

In order to give a Rakotch type characterization of the higher-order Hardy-Rogers-type contraction we introduce the following class of functions

Definition F.9 1

Let (X, d) be a metric space, $T : X \mapsto X$ be a mapping and let ρ and $\rho^\star$ be w-distances on X. We denote by Λ the family of functions $\varrho(x, y)$ satisfying the following conditions

(a) $\varrho(x, y) = \varrho(\rho^\star(x, y))$, where $\rho^\star(x, y) = \rho(x, Tx) + \rho(y, Ty) + \rho(x, Ty) + \rho(y, Tx) + \rho(x, y)$

(b) $0 \leq \varrho(\rho^\star) < 1$ for every $\rho^\star > 0$

(c) $\varrho(\rho^\star)$ is a monotonically decreasing function of $\rho^\star$

Definition F.10 1

Let (X, d) be a metric space and let ρ and $\rho^\star$ be w-distances on X. A map $T : X \mapsto X$ will be called a w-Rakotch higher-order Hardy-Rogers type contraction if there exists $\varrho(x, y) \in \Lambda$ such that $\rho(T^r x, T^r y) \leq \varrho(x, y)\rho^\star(x, y)$, for all $x, y \in X$ and any $r \in \mathbb{N}$

Remark F.11 1

If $\rho = d$ in the previous definition, then we say T is a Rakotch higher-order Hardy-Rogers type contraction

Remark F.12 1

If $\varrho(x, y) := Z\beta^r$ where $Z \geq 1$ and $\beta \in [0, \frac{1}{5})$ come from Proposition A.3, then for all $x, y \in X$, we have, $\rho(T^r x, T^r y) \leq Z\beta^r \rho^\star(x, y)$, for any $r \in \mathbb{N}$, and we will say T is a w-higher-order Hardy-Rogers type contraction

Remark F.13 1

If in Remark F.12, $\rho = d$, then T is a higher-order Hardy-Rogers type contraction

Remark F.14 1

Let $\varrho(x, y)$ be defined as in Remark F.12. Since $\varrho(x, y) < 1$, then $x \neq y$ implies $\rho(T^r x, T^r y) < \rho^\star(x, y)$, and we say T is a w-higher-order Hardy-Rogers type contractive mapping

Remark F.15 1

If in Remark F.14, $\rho = d$, then we say T is a higher-order Hardy-Rogers type contractive mapping

6.3 Main Results

> **Theorem F.1(b) 1**
>
> Let (X, d) be a complete metric space and let ρ and $\rho^\star$ be w-distances on X. Let $T : X \mapsto X$ be a w-Rakotch higher-order Hardy-Rogers type contraction. Then there exists a unique $z \in X$ such that $T^r z = z$. Further z satisfies $\rho(z, z) = 0$

> **Proof of Theorem F.1(b) 1**
>
> Let $x_0 \in X$ and define $x_n = T^{rn} x_0$ for $n \in \mathbb{N}$ and any $r \in \mathbb{N}$. Since T is a w-Rakotch higher-order Hardy-Rogers type contraction, we have, $\rho(x_n, x_{n+1}) = \rho(T^r x_{n-1}, T^r x_n) \leq \varrho(x_{n-1}, x_n) \rho^\star(x_{n-1}, x_n)$, and we notice that $\rho(x_n, x_{n+1}) \leq \varrho''(x_{n-1}, x_n) \rho(x_n, x_{n-1})$, where, $\varrho''(x_{n-1}, x_n) := \frac{3\varrho(x_{n-1}, x_n)}{1 - 2\varrho(x_{n-1}, x_n)}$. Clearly, ϱ'' is dependent on the w-distance $\rho^\star$ on X, $0 \leq \varrho''(\rho^\star) < 1$ for every $\rho^\star > 0$, and $\varrho''(\rho^\star)$ is a monotonically decreasing function of $\rho^\star$. So $\varrho'' \in \Lambda$. Thus,
>
> $$\rho(x_n, x_{n+1}) \leq \varrho''(x_{n-1}, x_n) \rho(x_n, x_{n-1}) \leq \cdots \leq \prod_{k=0}^{n-1} \varrho''(\rho(x_k, x_{k+1})) \rho(x_0, T^r x_0)$$
>
> Let $\epsilon_0 > 0$, if $\rho(x_k, x_{k+1}) \geq \epsilon_0$ for $k = 0, 1, \cdots, n-1$, then since ϱ'' is monotone, $\varrho''(\rho(x_k, x_{k+1})) \leq \varrho''(\epsilon_0)$, and hence, $\rho(x_n, x_{n+1}) \leq (\varrho'')^n(\epsilon_0) \rho(x_0, T^r x_0)$, but, $0 \leq (\varrho'')^n(\epsilon_0) < 1$, therefore by Lemma F.5, $\lim_{n \to \infty} \rho(x_n, x_{n+1}) = 0$. Now we show that $\{x_n\}$ is a Cauchy sequence in (X, d). For $m > 0$,
>
> $$\rho(x_n, x_{n+m}) \leq \prod_{k=0}^{n-1} \varrho''(\rho(x_k, x_{k+m})) \rho(x_0, T^r x_0)$$
>
> If $\rho(x_k, x_{k+m}) \geq \epsilon_0$ for any $\epsilon_0 > 0$ and $k = 0, 1, \cdots, n-1$, then, $\rho(x_n, x_{n+m}) \leq (\varrho'')^n(\epsilon_0) \rho(x_0, T^r x_0) \to 0$ as $n \to \infty$, and so by Lemma F.5, the sequence is Cauchy. Since (X, d) is complete, $\{x_n\}$ converges to some $z \in X$. Since $x_m \to z$ and $\rho(x_n, \cdot)$ is lower semi-continuous, $\rho(x_n, z) \leq \lim_{m \to \infty} \rho(x_n, x_m) \leq (\varrho'')^n(\epsilon_0) \rho(x_0, T^r x_0)$. Thus, $\lim_{n \to \infty} \rho(x_n, z) = 0$. On the other hand,
>
> $$\rho(x_n, T^r z) = \rho(T^r x_{n-1}, T^r z) \leq \varrho''(x_{n-1}, z) \rho(x_{n-1}, z) < \rho(x_{n-1}, z)$$
>
> So, $\lim_{n \to \infty} \rho(x_n, T^r z) = 0$, and hence by Lemma F.5, $T^r z = z$. Now,
>
> $$\rho(z, z) = \rho(T^r z, T^r z) \leq \varrho''(z, z) \rho(z, z) < \rho(z, z)$$
>
> So $\rho(z, z) = 0$. Finally if $y = T^r y$, then,
>
> $$\rho(z, y) = \rho(T^r z, T^r y) \leq \varrho''(z, y) \rho(z, y) < \rho(z, y)$$
>
> So $\rho(z, y) = 0$ and by Lemma F.5, $z = y$

> **Remark F.1 1**
>
> If $\rho = d$, (X, d) is a complete metric space, and $T : X \mapsto X$ is a Rakotch higher-order Hardy-Rogers type contraction, then Theorem F.1(b) still holds

Remark F.2 1

If (X,d) is a complete metric space and T is a w-higher-order Hardy-Rogers type contraction, then Theorem F.1(b) still holds. In particular we get a generalization of the higher-order Hardy-Rogers type mapping theorem [Theorem A.2]

Theorem F.3 1

Let (X,d) be a complete metric space, and let ρ and $\rho^\star$ be w-distances on X. Let $T : X \mapsto X$ be a mapping such that for some positive integer $m \in \mathbb{N}$, T^{rm} is a w-Rakotch higher-order Hardy-Rogers type contraction, for any $r \in \mathbb{N}$. Then there exists a unique $z \in X$ such that $T^r z = z$. Further z satisfies $\rho(z,z) = 0$.

Proof of Theorem F.3 1

Since for some positive integer $m \in \mathbb{N}$, T^{rm} is a w-Rakotch higher-order Hardy-Rogers type contraction, for any $r \in \mathbb{N}$, it follows that for every $x,y \in X$, $\rho(T^{rm}x, T^{rm}y) \leq \varrho''(x,y)\rho(x,y)$, where $\varrho''(x,y) := \frac{3\varrho(x,y)}{1-2\varrho(x,y)}$. Hence by Theorem F.1(b), there exist $z \in X$ such that $z = T^{rm}z$ for some $m \in \mathbb{N}$ and any $r \in \mathbb{N}$, and $T^r z = T^r(T^{rm}z) = T^{rm}(T^r z)$, thus $z = T^r z$. Further $\rho(z,z) = 0$ is clear.

Remark F.4(b) 1

Theorem F.3 still holds, if $\rho = d$ and for some positive integer $m \in \mathbb{N}$, T^{rm} is a higher-order Hardy-Rogers type contraction, for any $r \in \mathbb{N}$

Theorem F.5 1

Let ρ and $\rho^\star$ be w-distances associated with the metric space (X,d); let q be a w-distance associated with the metric space (X,v). Suppose $T : X \mapsto X$ is a mapping such that the following holds

(a) $\rho(x,y) \leq q(x,y)$

(b) (X,d) is complete

(c) $T : (X,v) \mapsto (X,v)$ is a w-Rakotch higher-order Hardy-Rogers type contraction, that is, there exists $\varrho(x,y) \in \Lambda$ such that $q(T^r x, T^r y) \leq \varrho(x,y) q^\star(x,y)$ for all $x,y \in X$, where $q^\star(x,y) := q(x,y) + q(x,Tx) + q(y,Ty) + q(x,Ty) + q(y,Tx)$

Then there exists $z \in X$ such that $T^r z = z$ for any $r \in \mathbb{N}$ and z satisfies $\rho(z,z) = 0$

Remark F.6 1

Notice that if $x_0 \in X$ and we define for $n \in \mathbb{N}$ and any $r \in \mathbb{N}$, $x_n = T^{rn} x_0$, then from (c), $\{x_n\}$ is Cauchy in (X,v). By (a) and Lemma F.5, $\{x_n\}$ is Cauchy in (X,d), and by (b) it converges. Therefore, the remaining aspect of the proof of the above theorem is similar to Theorem F.1(b)

Theorem F.7 1

Let $\epsilon > 0$ be given and let (X, d) be a complete ϵ-chainable metric space. If T is a mapping from X into itself satisfying $0 < d(x,y) < \epsilon$ implies $d(T^r x, T^r y) \leq \varrho(x,y) d^\star(x,y)$ for all $x, y \in X$, where $d^\star(x,y) := d(x,y) + d(x, Tx) + d(y, Ty) + d(x, Ty) + d(y, Tx)$, and $\varrho(x,y) \in \Lambda$. Then T has a unique $z \in X$ such that $z = T^r z$ for any $r \in \mathbb{N}$

Proof of Theorem F.7 1

Since (X, d) is ϵ-chainable for every $x, y \in X$, we define the function $\rho : X \times X \mapsto [0, \infty)$ as follows $p(x,y) = \inf\{\sum_{i=1}^n d(x_{i-1}, x_i) : \{x_0, x_1, \cdots, x_n\}$ is an ϵ-chain joining x and $y\}$. Note that ρ is a w-distance satisfying $d(x,y) \leq \rho(x,y)$. Given $x, y \in X$ and an ϵ-chain $\{x_0, x_1, \cdots, x_n\}$ with $x_0 = x$ and $x_n = y$ we have for $i = 1, \cdots, n$, $d(T^r x_{i-1}, T^r x_i) \leq \varrho''(d(x_{i-1}, x_i)) d(x_{i-1}, x_i)$, where $\varrho''(d(x_{i-1}, x_i)) := \frac{3\varrho(d(x_{i-1}, x_i))}{1 - 2\varrho(d(x_{i-1}, x_i))}$. Therefore,
$d(T^r x_{i-1}, T^r x_i) \leq \varrho''(\epsilon)\epsilon < \epsilon$. Hence $T^r x_0, \cdots, T^r x_n$ is an ϵ-chain joining $T^r x$ and $T^r y$. Now, $\rho(T^r x, T^r y) \leq \sum_{i=1}^n d(T^r x_{i-1}, T^r x_i) \leq \sum_{i=1}^n \varrho''(d(x_{i-1}, x_i)) d(x_{i-1}, x_i)$. Since $\{x_0, \cdots, x_n\}$ is an arbitrary ϵ-chain, we have, $\rho(T^r x, T^r y) \leq \varrho''(x,y) \rho(x,y)$, hence by Theorem F.1(b), T has a unique fixed point $z \in X$ such that $z = T^r z$ for any $r \in \mathbb{N}$

Theorem F.8 1

Let $\epsilon > 0$ be given, and let (X, d) be a complete ϵ-chainable metric space. If T is a mapping from X into itself satisfying for some $m \in \mathbb{N}$ and any $r \in \mathbb{N}$, the condition, $d(x,y) < \epsilon$ implies $d(T^{rm} x, T^{rm} y) \leq \varrho(x,y) d^\star(x,y)$, for every $x, y \in X$, where $\varrho(x,y) \in \Lambda$ and $d^\star(x,y) := d(x,y) + d(x, Tx) + d(y, Ty) + d(x, Ty) + d(y, Tx)$, then there exist a unique $z \in X$ such that $z = T^r z$ for any $r \in \mathbb{N}$

Proof of Theorem F.8 1

As in the previous theorem, we define the function $\rho : X \times X \mapsto [0, \infty)$ as follows $p(x,y) = \inf\{\sum_{i=1}^n d(x_{i-1}, x_i) : \{x_0, x_1, \cdots, x_n\}$ is an ϵ-chain joining x and $y\}$, and note that ρ is a w-distance satisfying $d(x,y) \leq \rho(x,y)$. Further we observe that for some $m \in \mathbb{N}$ and any $r \in \mathbb{N}$, T^{rm} satisfies the condition $\rho(T^{rm} x, T^{rm} y) \leq \varrho(x,y) \rho(x,y)$ for some $\varrho(x,y) \in \Lambda$ and all $x, y \in X$. Thus by the previous theorem, there exist a unique $z \in X$ such that $z = T^{rm} z$. It follows that there exist a unique $z \in X$ such that $z = T^r z$

Theorem F.9 1

Let (X, d) be a noncomplete metric space and ρ and $\rho^\star$ be w-distances on X. Let $T : X \mapsto X$ be a w-Rakotch higher-order Hardy-Rogers type contraction and suppose there exist a point $u \in X$ such that $\theta(u) = \inf\{\theta(x) : x \in X\}$, where $\theta(x) = \rho(x, Tx)$ for all $x \in X$, and any $r \in \mathbb{N}$. Then $u = T^r u$ for any $r \in \mathbb{N}$.

Proof of Theorem F.9 1

If $u = T^r u$, then we are done. We suppose $u \neq T^r u$. Since T is a w-Rakotch higher-order Hardy-Rogers type contraction, there exist $\varrho(x,y) \in \Lambda$ such that $\rho(T^r x, T^r y) \leq \varrho(x,y)\rho^\star(x,y)$, where $\rho^\star(x,y)$ comes from Definition F.9. Hence it follows that $\rho(T^r x, T^r y) \leq \varrho''(x,y)\rho(x,y)$, where, $\varrho''(x,y) := \frac{3\varrho(x,y)}{1-2\varrho(x,y)}$. Hence, we have,

$$\theta(T^r u) = \rho(T^r u, T^{2r} u) \leq \varrho''(u, T^r u)\rho(u, T^r u) < \rho(u, T^r u) = \theta(u)$$

, which is a contradiction.

6.4 Exercises

Exercise F.1 1

Let (X, d) be a metric space. Prove that if $T : X \mapsto X$ is a Reich type mapping (Exercise A.1), then T has a unique fixed point provided that X is complete.

Exercise F.2 1

Verify Example F.3

Exercise F.3 1

In Definition F.9, replace $\rho^\star$ with $\rho^{\star\star} := \rho(x, Tx) + \rho(y, Ty) + \rho(x, y)$, and deduce from this new class of functions what it will mean for a map $T : X \mapsto X$ to be a

(a) w-Rakotch higher-order Reich type contraction mapping

(b) Rakotch higher-order Reich type contraction mapping

(c) w-higher-order Reich type contraction mapping

(d) higher-order Reich type contraction mapping

(e) w-higher-order Reich type contractive mapping

(f) higher-order Reich type contractive mapping

Exercise F.4 1

Let (X, d) be a metric space and ρ and $\rho^{\star\star}$ be w-distances on X, where $\rho^{\star\star}$ is given by the previous exercise. Prove the following

(a) If $T : X \mapsto X$ is a w-Rakotch higher-order Reich type contraction. Then there exists a unique $z \in X$ such that $T^r z = z$. Further z satisfies $\rho(z, z) = 0$.

(b) If $T : X \mapsto X$ is a mapping such that for some positive integer $m \in \mathbb{N}$, T^{rm} is a w-Rakotch higher-order Reich type contraction, for any $r \in \mathbb{N}$, then there exists a unique $z \in X$ such that $T^r z = z$. Further z satisfies $\rho(z, z) = 0$.

> **Exercise F.5 1**
>
> Let (X, d) be a non-complete metric space and ρ and $\rho^{\star\star}$ be w-distances on X, where $\rho^{\star\star}$ is given by Exercise F.3. Let $T : X \mapsto X$ be a w-Rakotch higher-order Reich type contraction and suppose there exist a point $u \in X$ such that $\theta(u) = \inf\{\theta(x) : x \in X\}$, where $\theta(x) = \rho(x, T^r x)$ for all $x \in X$, and any $r \in \mathbb{N}$. Then $u = T^r u$ for any $r \in \mathbb{N}$.

6.5 References

(1) Rakotch, E. (1962) "A note on contractive mappings", Proc. A.M.S. 13: 459–465

(2) Kada, O.; Suzuki, T.; Takahashi, W. (1996) "Nonconvex minimization theorems and fixed point theorems in complete metric spaces", Math. Japon. 44: 381–391

(3) Canad. Math. Bull. Vol. 16 (2), 1973

www.ingramcontent.com/pod-product-compliance
Lightning Source LLC
Chambersburg PA
CBHW051105180526
45172CB00002B/782